Psybersecurity

Human Factors of Cyber Defence

Edited by
Oliver Guidetti, Mohiuddin Ahmed,
and Craig Speelman

CRC Press
Taylor & Francis Group
Boca Raton London New York

CRC Press is an imprint of the
Taylor & Francis Group, an **informa** business

Designed cover image: © Shutterstock

First edition published 2025
by CRC Press
2385 NW Executive Center Drive, Suite 320, Boca Raton FL 33431

and by CRC Press
4 Park Square, Milton Park, Abingdon, Oxon, OX14 4RN

CRC Press is an imprint of Taylor & Francis Group, LLC

ISBN: 978-1-032-66483-5 (hbk)
ISBN: 978-1-032-66484-2 (pbk)
ISBN: 978-1-032-66485-9 (ebk)

DOI: 10.1201/9781032664859

Typeset in Sabon
by Apex CoVantage, LLC

Dedicated to the victims of cybersecurity breaches. I hope this book contributes towards a world where our information is far more secure, Professor Craig Speelman.

Dedicated to my sisters-in-law: Farzana Nipun and Dr Farzana Islam, Dr Mohiuddin Ahmed.

Dedicated to my son Oakley, Dr Oliver Guidetti.

Contents

About the editors

Oliver Guidetti is a postdoctoral researcher specialising in the human factors of cybersecurity. He has an extensive background working on applied socio-technical projects that examine the interplay between the human brain and technology. His PhD research focused on measuring and mitigating vigilance decrement in cyber network defence tasks. Guidetti has published numerous research papers on topics such as cyber vigilance tasks, the neuropsychology of network defence operators, human resource policy in relation to neurotechnology, and the effects of cognitive load on the performance of cyber first responders. Through his work, he hopes to contribute to our increasingly digital world's socio-technical security and resilience.

Mohiuddin Ahmed is Senior Lecturer of Computing and Security Discipline at the School of Science, Edith Cowan University, Australia. He earned his PhD in computer science from the University of New South Wales in 2016. He is a member of the Association for Computing Machinery (ACM), the Australian Information Security Association (AISA), a senior member of the Institute of Electrical and Electronics Engineers (IEEE), a Certified Professional of the Australian Computer Society (ACS), and a Fellow of the Higher Education Academy, UK. He is also an ACM Distinguished Speaker and delivered several keynotes and invited talks on cybersecurity. He has been educating the next generation of cyber leaders and researching to disrupt the cybercrime ecosystem. His research is focused on ensuring national security and safeguarding critical infrastructures from cyber terrorists. He secured several external and internal grants worth more than AU$1.7 million and has been collaborating with academia and industry.

Craig Speelman has been Professor at the School of Psychology and Social Science, Edith Cowan University, since 2008. He was Head of the School during 2002–2011. Speelman conducts research in the general area of cognitive psychology, but much of his research has focused on the topic of skill acquisition and transfer. He is the author of *Beyond the Learning*

Curve: The Construction of Mind and several book chapters. He has edited three research volumes and also authored papers published in highly ranked journals, including *Journal of Experimental Psychology: LMC*, *Nature Scientific Reports*, *Frontiers in Psychology*, *The American Journal of Geriatric Psychiatry*, and *PLOS ONE*. He has received funding from the Australian Research Council, Health Workforce Australia, and the Association of Independent Schools (AISWA). He has served on editorial boards of journals and was the editor-in-chief of the *GSTF Journal of Psychology*. He has served on the organising committees for national and international psychology conferences. Speelman has applied this expertise to developing a training programme for acquiring skin cancer detection skills, the development of basic arithmetic skills, where he produced the commercial app Numbeat, and psychological aspects of cybersecurity.

Contributors

Tooba Aamir
CSIRO
Clayton
Victoria

Mohiuddin Ahmed
Edith Cowan University
Joondalup
Western Australia

Josh Armitage
Cognizant, Docklands
Victoria
Australia

Barry L. Bentley
Cardiff School of Technologies
Cardiff Metropolitan University
Cardiff, Wales, UK

Fiona Carroll
Cardiff School of Technologies
Cardiff Metropolitan University
Cardiff, Wales, UK

Osmar Couto
Edith Cowan University
Joondalup
Western Australia

Anna J. Griffin
Edith Cowan University
Joondalup
Western Australia

Marthie Grobler
CSIRO
Clayton
Victoria

Alexander E. Grojek
Edith Cowan University
Joondalup
Western Australia

Oliver Guidetti
Edith Cowan University
Joondalup
Western Australia

David J. Holmes
Edith Cowan University
Joondalup
Western Australia

Ahmed Ibrahim
Edith Cowan University
Joondalup
Western Australia

Nicola F. Johnson
Edith Cowan University
Joondalup
Western Australia

Marnie McKee
Edith Cowan University
Joondalup
Western Australia

Avisen Moonsammy
Deloitte, Edith Cowan University
Joondalup
Western Australia

Bazlur Rashid
Edith Cowan University
Joondalup
Western Australia

Leslie F. Sikos
Edith Cowan University
Joondalup
Western Australia

Craig Speelman
Edith Cowan University
Joondalup
Western Australia

Adrian Wood
Edith Cowan University
Joondalup
Western Australia

Shasha Yu
Clark University
Worcester
Massachusetts

Chapter 1

Integrating human factors and systemic resilience

An interdisciplinary approach to cybersecurity in critical infrastructures and utilities

Avisen Moonsammy, Mohiuddin Ahmed, Oliver Guidetti, and Bazlur Rashid

1.1 INTRODUCTION: THE CONVERGENCE OF DISCIPLINES IN CYBERSECURITY

In the ever-evolving digital landscape, critical infrastructures and utilities face a myriad of cybersecurity threats. As Clark and Hakim (2017) identified, the challenge is no longer confined to the virtual realm; it extends to the physical world as cyber-physical systems become increasingly integral to critical operations. Despite the growing recognition of these cyber-physical threats (Cohen, 2010; Gunduz & Das, 2020), one area remains underexplored: the human element in cybersecurity. As Bates (2017) and Boyce et al. (2011) have pointed out, human performance significantly influences cybersecurity. Nevertheless, current cybersecurity strategies often fail to incorporate human factors fully, leaving utilities and other critical infrastructures vulnerable to human-induced risks (Hellström, 2007). The growing complexity of cyberthreats necessitates an interdisciplinary approach to cybersecurity that bridges the gap between psychology, systems theory, and cybersecurity (Boyce et al., 2011). This approach recognises that the key to enhancing cyber defence lies in technological advancement and understanding and managing human behaviour (Bates, 2017). Furthermore, as Von Solms and Van Niekerk (2013) and Panguluri et al. (2017) have shown, cybersecurity culture in organisations plays a crucial role in shaping this behaviour. The significance of this study is underscored by the urgency of addressing systemic vulnerabilities (Hellström, 2007) and enhancing cyber resilience (Kott & Linkov, 2019), particularly in an era where AI is poised to play a significant role in cybersecurity (Takahashi et al., 2010).

DOI: 10.1201/9781032664859-1

1.1.1 Statement of the problem

The digitalisation of our society, especially in critical infrastructures like utilities and health, has presented numerous benefits, such as increased efficiency, improved reliability, and unprecedented convenience. However, this reliance on digital systems also exposes these infrastructures to cybersecurity threats, which, if not correctly managed, could result in severe consequences, including service disruptions, loss of lives, financial losses, and potential threats to national security. Cybersecurity in critical infrastructures involves complex and interconnected issues, such as the evolving nature of cyberthreats, the intertwined roles of technology and humans in vulnerability and defence, the importance of systemic resilience, and the potential of emerging technologies like AI in enhancing cybersecurity measures. The human factor represents a significant yet often underestimated element of cybersecurity. Employees' behaviours, organisational culture, and leadership commitment are crucial in shaping an organisation's cybersecurity posture.

Furthermore, traditional approaches to cybersecurity that focus on individual aspects, such as information security or technical defences, have been found insufficient in the face of multifaceted and sophisticated cyberthreats. Integrated cybersecurity frameworks, which encompass multiple layers of defence and consider human and technical factors, are increasingly recognised as crucial. Nevertheless, understanding and implementing such integrated approaches remain challenging for many organisations. Moreover, while systemic resilience – the ability of a system to withstand and adapt to disturbances – is critical in cybersecurity, it is a concept that is not yet fully understood or operationalised in many critical infrastructures. However, AI holds significant potential for enhancing cybersecurity, and harnessing this potential effectively requires further exploration.

In light of these complexities and challenges, this chapter seeks to comprehensively examine cybersecurity in critical infrastructures, focusing on the human factor, integrated cybersecurity frameworks, systemic resilience, and the role of AI. It aims to contribute to the understanding and practice of cybersecurity, ultimately helping safeguard our critical infrastructures against current and future cyberthreats.

1.1.2 Significance of this study

This study is relevant to cybersecurity, especially critical infrastructures, and is essential from multiple perspectives (Figure 1.1).

1. *Comprehensive Understanding of Cybersecurity:* By adopting an integrated view of cybersecurity, this study enhances the understanding of the multifaceted and interdependent nature of cybersecurity issues. It goes beyond technical defence and includes human and cultural elements, systemic resilience, and emerging technologies. This contributes to a more comprehensive cybersecurity.

01	02	03	04	05
Comprehensive Understanding of Cybersecurity	Insights into Human-Related Risks	Promotion of Integrated Cybersecurity Frameworks	Exploration of Systemic Resilience	Harnessing the Potential of AI

Figure 1.1 Study Significance.

2. *Insights Into Human-Related Risks:* The study makes a significant contribution to discussions on human factors in cybersecurity, a topic that has often been overshadowed by a focus on technological aspects. By shedding light on human-related risks, this research underscores the importance of shaping organisational culture and training programmes to mitigate these risks.
3. *Promotion of Integrated Cybersecurity Frameworks:* The study offers practical insights into how these frameworks can be developed and implemented by advocating for integrated cybersecurity frameworks. It can guide organisations, especially those operating critical infrastructures, in designing cybersecurity strategies to address various threats and vulnerabilities.
4. *Exploration of Systemic Resilience:* The research brings systemic resilience to the fore in cybersecurity discussion. This is a relatively underexplored yet vital aspect of cybersecurity, and the study's emphasis on systemic resilience can help organisations strengthen their capacity to withstand and recover from cyberattacks.
5. *Harnessing the Potential of AI:* The study also paves the way for further discussions and investigations into the use of AI in cybersecurity. The insights it provides can assist in driving forward the practical and responsible application of AI technologies in enhancing cybersecurity defences.

This study can potentially inform and influence the cybersecurity strategies and practices of organisations operating critical infrastructures by focusing on these aspects. The insights generated from this research can play a vital role in bolstering cybersecurity measures, thereby safeguarding critical services, economies, and societies.

1.1.3 Research objectives

The primary objective of this chapter is to examine the intersection of human factors and systemic vulnerabilities in cybersecurity, focusing on utilities and other critical infrastructures. In doing so, this chapter aims at the following:

1. Explore the role of human behaviour in cybersecurity, including its influence on systemic vulnerabilities and potential as a focus for cyber defence (Boyce et al., 2011; Kovács & Nyikes, 2020)

2. Investigate the impact of cybersecurity culture on human behaviour and system resilience (Panguluri et al., 2017; Von Solms & Van Niekerk, 2013)
3. Develop an integrated framework for cybersecurity that effectively addresses both human factors and systemic vulnerabilities (Cormier & Ng, 2020)
4. Evaluate the potential of AI in mitigating human-related risks (Takahashi et al., 2010)
5. Offer practical recommendations for enhancing cyber defence through a comprehensive understanding of the human element (Bates, 2017; Gunduz & Das, 2020)

1.1.4 Chapter overview

This chapter is organised into nine sections, each exploring a different aspect of the human element in cybersecurity. Following this introduction, Section 1.2 reviews the concept of cyber-physical security and its importance in critical infrastructures. Section 1.3 examines systemic vulnerabilities in these infrastructures and how they can be exploited in cyberattacks. Section 1.4 delves into the human factor in cybersecurity, discussing human errors, social engineering, and insider threats. Section 1.5 discusses the role of culture in cybersecurity, while Section 1.6 proposes an integrated approach to cybersecurity. Section 1.7 explores the concept of systemic resilience in cybersecurity. Finally, Section 1.8 looks at the future of cybersecurity, particularly the role of AI, before the chapter concludes with a summary of insights and practical recommendations in Section 1.9. As such, this chapter seeks to contribute to the existing literature by offering an in-depth exploration of the human element in cybersecurity, emphasising utilities and other critical infrastructures.

1.2 CYBER-PHYSICAL SECURITY AND CRITICAL INFRASTRUCTURE: THE INDIVISIBLE DUO

1.2.1 Definition of key terms

Cyber-Physical Systems (CPS): Cyber-physical systems represent the convergence of computational resources and physical processes. CPS leverages computing, communication, and control technologies to interact with the physical world in real time, generating tangible outcomes based on digital instructions (Gunduz & Das, 2020). As shown in Figure 1.2, CPS encompasses everything from Operational Technology (OT) to the Industrial Internet of Things (IIoT), Smart Cities, Smart Buildings, Industry 4.0, etc. In the context of this chapter, CPS is central to understanding how digital security measures directly influence the safe operation of utilities and other critical infrastructures.

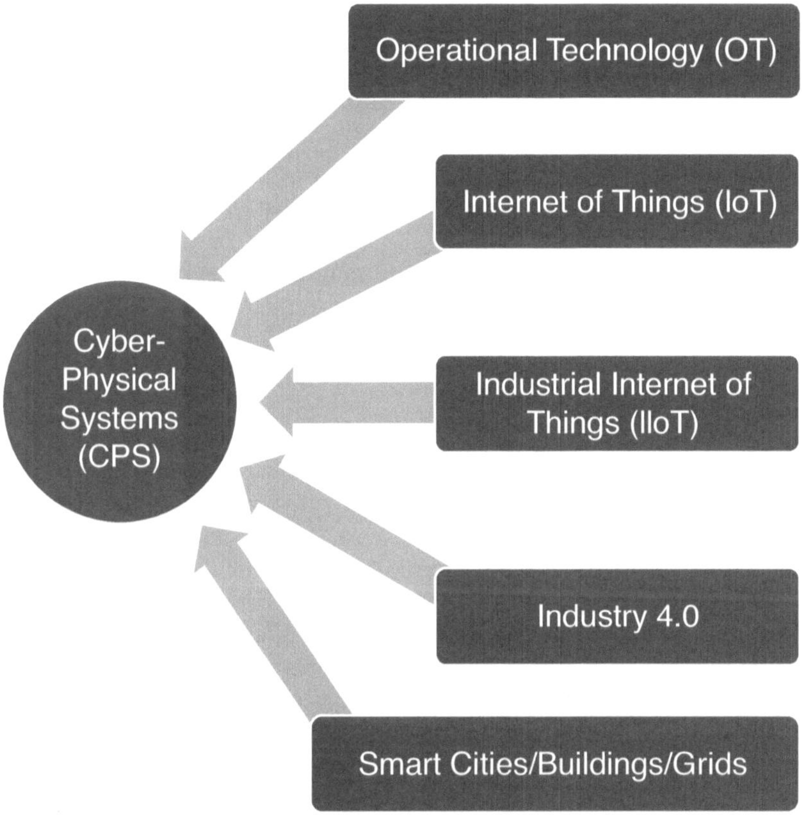

Figure 1.2 Cyber-Physical System (CPS) Components.

Critical infrastructures are systems, networks, and physical and cyber–based assets that society heavily relies upon. They encompass vital sectors such as water and wastewater utilities, power grids, and transportation systems. The incapacitation or destruction of such infrastructures would significantly impact security, national economic security, public health, or safety. As highlighted by Cohen (2010), safeguarding these assets from cyberthreats is paramount to maintaining societal order and well-being.

Cyber-physical security pertains to measures implemented to shield cyber-physical systems from threats that could cause physical or virtual damage. It involves preserving the integrity of a system's computational and physical elements. Clark and Hakim (2017) advocate for a multifaceted, integrated approach to ensuring the security of these interconnected systems.

These definitions provide a foundational understanding for this chapter's exploration of the human factors impacting cybersecurity in critical infrastructures. The discussion on systemic vulnerabilities and their contribution to cybersecurity threats becomes more relevant by appreciating the interconnectedness and vulnerability of CPS and critical infrastructures. Examining the impact of systemic vulnerabilities on cybersecurity threats highlights the significance of a company's cybersecurity culture in mitigating risks. People play a critical role in operating and protecting these systems. Finally, the definitions also serve as a stepping stone to exploring the future of cybersecurity. As technology advances, AI's role in mitigating threats, streamlining responses, and enhancing the security of CPS and critical infrastructures becomes increasingly significant, signifying an exciting frontier in cybersecurity.

1.2.2 Evolution and importance of cyber-physical systems

The rise of CPS has mirrored advancements in computing and communication technologies. These developments have led to increasing digitisation and interconnection of systems, blurring the boundaries between the physical and cyber worlds. This evolution has given rise to CPS, capable of sophisticated interactions with the physical world, driving efficiencies and opportunities in numerous sectors (Gunduz & Das, 2020). The importance of CPS in today's society cannot be overstated. Their integration into critical sectors such as utilities, healthcare, transportation, and manufacturing has made them essential components of our everyday lives. CPS streamline processes, increase efficiency, and offer unparalleled adaptability in managing complex, real-time operations. However, the prevalence of these systems has also drawn the attention of cybercriminals who seek to exploit their vulnerabilities for malicious intent (Clark & Hakim, 2017). This highlights the first objective of this chapter, which is to provide an in-depth understanding of the human factors affecting cybersecurity in critical infrastructures. As CPS continues to permeate essential sectors, understanding how human operators interact with these systems and where potential vulnerabilities exist becomes essential for ensuring their secure operation.

Additionally, the evolution and importance of CPS underscore the need to discuss systemic vulnerabilities and their contribution to cybersecurity threats. CPS are characterised by their interconnectivity, which, while providing numerous benefits, also presents a series of complex vulnerabilities that can be exploited in cyberattacks (Hellström, 2007). Furthermore, the pervasive nature of CPS across various sectors highlights the crucial role of an organisation's cybersecurity culture in risk mitigation. Given the complexity and criticality of these systems, fostering a solid cybersecurity culture can help minimise human errors, encourage proactive protection measures, and enhance the overall security posture of organisations operating these systems (Panguluri et al., 2017; Von Solms & Van Niekerk, 2013). Given the complexity of these systems, this chapter proposes an integrated cybersecurity framework. This will incorporate risk analyses and hazard management strategies adapted to the unique characteristics of CPS, fostering systemic resilience against potential cyber-physical attacks (Cormier & Ng, 2020; Kott & Linkov, 2019). Finally, considering the fast-paced evolution of CPS and the challenges they present, the role of AI has become increasingly crucial in cybersecurity. AI's potential to automate detection, improve response times, and predict threats can significantly enhance the security of CPS, pointing to an exciting direction for the future of cybersecurity (Takahashi et al., 2010).

1.2.3 Specific challenges in cyber-physical security

Cyber-physical security presents unique challenges due to CPS's convergence of cyber and physical domains. One such challenge is the real-world ramifications of cyberthreats. In CPS, a cyberattack can manifest physical consequences, thus extending the threat landscape beyond data breaches or system disruptions (Gunduz & Das, 2020). For instance, a cyberattack on a water treatment plant could manipulate control systems, leading to water contamination or service disruption. Such attacks can have far-reaching consequences on public health and safety, highlighting the tangible risks of cyberthreats in CPS (Gunduz & Das, 2020).

Furthermore, the systemic interconnectedness of CPS introduces another layer of complexity to cybersecurity. Systemic vulnerabilities – weaknesses in one part of the system that could be exploited to compromise the entire system – make defending CPS an intricate task (Hellström, 2007). An attacker can exploit a single vulnerability point to cause severe damage to an entire CPS system, showing the multiplying effect of systemic vulnerabilities. The human element further complicates the cybersecurity landscape. Human operators, whether they interact directly with CPS or indirectly through decision-making roles, can significantly influence the security of these systems (Bates, 2017; Boyce et al., 2011). Their actions, both intentional (as in the case of insider threats) and accidental (such as errors due to lack of training or awareness), can introduce vulnerabilities or exacerbate existing ones (Kovács & Nyikes, 2020). It is crucial to comprehend and handle the human elements involved in cybersecurity. This highlights the significance of understanding and managing human factors in cybersecurity. Addressing these challenges necessitates an integrated, multidisciplinary approach to cyber-physical security. Strategies must incorporate technical defences, human factor considerations, and organisational strategies like risk management and resilience-building (Cormier & Ng, 2020).

1.3 SYSTEMIC VULNERABILITIES AND CYBERSECURITY THREATS

1.3.1 Systemic vulnerabilities: an overview

Systemic vulnerabilities pertain to weaknesses inherent within a system or network that threat actors could capitalise upon to undermine its integrity (Hellström, 2007). They are not isolated incidents; instead, they form part of the interconnected web of a system's architecture. In the context of critical infrastructures, these vulnerabilities often emerge from the intricate complexity and interconnectivity intrinsic to CPS (Gunduz & Das, 2020). These vulnerabilities can take various forms. Architectural vulnerabilities, for instance, arise from designing and implementing a system or network (Cormier & Ng, 2020). They can also be linked to outdated or unsupported systems that no longer receive security patches, exposing them to known and emerging threats (Clark & Hakim, 2017). Weak authentication mechanisms, which can enable unauthorised access, form another significant category of systemic vulnerabilities (Von Solms & Van Niekerk, 2013).

Additionally, human errors contribute significantly to systemic vulnerabilities (Boyce et al., 2011). This can be misconfigurations, weak passwords, clicking on malicious links, or not adhering to cybersecurity best practices. These human-induced vulnerabilities underscore the need for cybersecurity education and awareness as part of an organisation's defence strategy (Kovács & Nyikes, 2020).

Understanding and addressing these systemic vulnerabilities is paramount to enhancing the security of CPS. This aligns with the second objective of this chapter, which seeks to discuss systemic vulnerabilities and their contribution to cybersecurity threats (Hellström, 2007). By doing so, the aim is to inform the development of robust, holistic cybersecurity strategies that account for the multifaceted nature of systemic risks in CPS. However, addressing systemic vulnerabilities alone is not enough. This chapter also explores the human factors affecting cybersecurity in critical infrastructures and highlights the importance of an organisation's cybersecurity culture in mitigating these vulnerabilities (Bates, 2017; Panguluri et al., 2017).

1.3.2 The interface between systemic vulnerabilities and cybersecurity

Systemic vulnerabilities are intimately linked with the concept of cybersecurity. Owing to the interconnected nature of CPS, a vulnerability in one section of a system can propagate, enabling an attacker to compromise other parts of the system, sometimes with catastrophic results (Clark & Hakim, 2017; Hellström, 2007). This domino effect indicates how systemic vulnerabilities can escalate from isolated incidents to pervasive threats. For instance, consider a utility infrastructure employing an Industrial Control

System (ICS). In such a scenario, a security weakness within the ICS could be exploited to manipulate physical processes, resulting in service disruptions or even physical damage (Gunduz & Das, 2020). This example underlines how cybersecurity threats in CPS can transcend the digital realm and induce physical consequences.

Moreover, human elements introduce systemic vulnerabilities in cybersecurity. Human errors, such as mishandling sensitive data, misconfigurations, or susceptibility to phishing attacks, can contribute to systemic vulnerabilities (Bates, 2017). Lack of adequate training can result in staff not adhering to cybersecurity best practices. At the same time, malicious insiders pose threats that are particularly difficult to defend against due to their legitimate access to systems (Boyce et al., 201; Kovács & Nyikes, 2020). Addressing these vulnerabilities and comprehending their interplay with cybersecurity is crucial.

1.3.3 Cyberthreats leveraging systemic weaknesses

The landscape of cyberthreats is a vast expanse populated by diverse actors. These actors extend from individual hackers with potentially limited, albeit still dangerous, resources to organised crime groups who orchestrate sophisticated attacks. At the most severe end of the spectrum are nation-state actors with advanced capabilities and vast resources, enabling them to execute highly damaging attacks (Cohen, 2010). Moreover, the types of threats these actors can deploy are numerous and varied. They encompass malware, which can disrupt system operations or exfiltrate data, and ransomware, which lock users out of their systems until a ransom is paid. Another significant threat is the Advanced Persistent Threat (APT), a stealthy and continuous computer hacking process often orchestrated by nation-states to steal information or disrupt operations. These threats present unique risks and require specific mitigation strategies. Recognising and understanding these threats is critical in countering systemic vulnerabilities in cybersecurity.

1.4 THE HUMAN ELEMENT IN CYBERSECURITY: THE WEAKEST LINK

1.4.1 Human factors in cybersecurity: an overview

The role of the human element in cybersecurity is pivotal and cannot be understated. The security of cyber-physical systems depends not only on robust technological and architectural systems but also on how humans interact with and manage these systems (Bates, 2017). Humans can sometimes be the weakest link, as their actions can introduce vulnerabilities into the system (Boyce et al., 2011). Critical infrastructure systems are susceptible to human error due to their complexity and interconnected nature.

A seemingly minor mistake can trigger a chain of events, leading to significant damage or disruptions. These systems, essential for the functioning of a society, are made more vulnerable by their dependence on human operators, who are susceptible to making errors or being targeted by threat actors (Hellström, 2007). This understanding of human factors in cybersecurity aligns closely with the purposes of this chapter, which aims to provide an in-depth analysis of how human actions and behaviours impact the security of critical infrastructures. By acknowledging humans' role in creating or exacerbating systemic vulnerabilities, we can develop more effective strategies to address these vulnerabilities (Bates, 2017; Boyce et al., 2011).

The chapter also emphasises the importance of fostering a solid cybersecurity culture within organisations to mitigate risks. By training employees in cybersecurity best practices and encouraging a culture that values security, organisations can minimise human errors that may otherwise lead to breaches or other security incidents (Panguluri et al., 2017). Additionally, the proposed cybersecurity framework considers not just technical measures but also the management of human factors. By incorporating elements such as risk analysis, hazard management, and systemic resilience, the framework aims to provide a holistic solution to cybersecurity challenges in critical infrastructure (Cormier & Ng, 2020; Kott & Linkov, 2019).

1.4.2 Human errors: a breach in cyber defence

Human errors are a significant cause of security incidents, accounting for many cybersecurity breaches (Bates, 2017). These errors can take numerous forms, from simple mistakes such as misconfiguring security settings to more complex issues like failing to follow established security protocols (Boyce et al., 2011). Human errors, unintentional or deliberate, can potentially provide an open door for threat actors to gain unauthorised access to systems and carry out malicious activities (Kovács & Nyikes, 2020). Various factors contribute to these human errors. Stress, cognitive overload, lack of proper training, and a poor understanding of cybersecurity protocols all can increase the likelihood of mistakes. These factors are not only a concern for the individuals making the errors but also pose a systemic risk to the entire organisation, mainly when the errors occur in critical infrastructure systems (Bates, 2017; Boyce et al., 2011). We aim to understand better the human factors affecting cybersecurity in critical infrastructures. Understanding the root causes of human error can help formulate strategies to prevent such errors, mitigate their impacts, and reveal the need for a robust cybersecurity culture within organisations (Bates, 2017; Boyce et al., 2011). Training and education can help reduce human errors by increasing employees' awareness of the potential consequences of their actions and equipping them with the knowledge and skills to follow security protocols effectively (Panguluri et al., 2017).

1.4.3 The psychology behind social engineering

Social engineering represents a unique and highly effective class of cyberthreats that exploit the human element rather than technical vulnerabilities (Boyce et al., 2011). A deep understanding of human psychology and behavioural tendencies is central to social engineering attacks. By manipulating these tendencies, cybercriminals deceive individuals into performing actions compromising security, such as disclosing sensitive information or performing unauthorised transactions (Boyce et al., 2011). The human element is often considered the weakest link in the cybersecurity chain, making it the target of social engineering. It plays on natural human instincts and emotions, including trust, curiosity, fear, and a desire to be helpful. It also exploits social norms and relationships, such as authority and reciprocation, which can lead individuals to act against their better judgment (Boyce et al., 2011). For example, an attacker may pose as a superior or trusted colleague, leveraging the authority relationship to convince the victim to disclose sensitive data or execute a malicious action. Alternatively, an attacker may try to get the victim to perform a specific task, such as providing login credentials, by offering a favour or assistance and expecting the victim to reciprocate (Boyce et al., 2011). Understanding the psychology behind social engineering is critical in designing effective countermeasures. By studying the psychological principles of these attacks, organisations can better equip their employees with the knowledge and awareness to detect and avoid such threats (Boyce et al., 2011; Panguluri et al., 2017).

Moreover, comprehending human factors contributes to an effective cybersecurity culture. When employees understand how they might be manipulated, they are less likely to fall victim to such tactics. Consequently, regular training that focuses on the psychological aspects of social engineering can be crucial in building a resilient cybersecurity culture (Panguluri et al., 2017; Von Solms & Van Niekerk, 2013). Furthermore, integrating this understanding of human psychology into an organisation's cybersecurity framework can contribute to systemic resilience. By considering the technical and human elements of cybersecurity, organisations can build robust defences that address the full scope of potential threats (Cormier & Ng, 2020; Kott & Linkov, 2019). Finally, emerging technologies like AI may also significantly combat social engineering. By analysing behaviour and communication patterns, AI systems could identify and flag suspicious activities that may indicate a social engineering attempt, thereby providing another layer of defence against these psychologically driven cyberthreats (Takahashi et al., 2010).

1.4.4 Insider threats: a hidden menace

Insider threats are a significant and often under-recognised cybersecurity concern. Such threats are typically posed by individuals with authorised access to an organisation's systems, such as employees, contractors, or business partners

(Kovács & Nyikes, 2020). These individuals, whether acting out of malicious intent, negligence, or manipulation by external parties, can inflict considerable damage due to their privileged access and knowledge of the organisation's operations (Bates, 2017; Kovács & Nyikes, 2020). Insider threats are uniquely challenging due to their diverse origins. They can stem from disgruntled employees seeking retaliation, careless staff making unintentional mistakes or even spies who have intentionally infiltrated the organisation (Kovács & Nyikes, 2020). Such threats may also involve complex psychological and interpersonal dynamics, making them difficult to predict and detect (Bates, 2017). A daunting aspect of insider threats is their potential to bypass many traditional security controls. Insiders, especially IT-related ones, often deeply understand the organisation's network and security practices. This knowledge allows them to evade detection tools, manipulate systems, and even erase evidence of their activities (Kovács & Nyikes, 2020).

Addressing insider threats requires delicate handling due to potential legal and ethical implications. Strategies must balance security requirements with respect for individual privacy and legal boundaries. If handled improperly, reactive measures can harm morale and exacerbate mistrust within the organisation (Kovács & Nyikes, 2020). The growing recognition of the importance of insider threats has led to the call for a comprehensive approach to their management. Such an approach should be multifaceted, combining technical measures, policy frameworks, and human resource strategies (Panguluri et al., 2017). From a technical perspective, this may involve employing advanced user behaviour analytics to detect abnormal patterns and implementing stricter access controls (Kott & Linkov, 2019; Takahashi et al., 2010). Organisational policies should clearly define acceptable and unacceptable behaviours and guide the handling of sensitive information (Panguluri et al., 2017). The aspect of human resources is critical in minimising insider threats. Regular training programmes, comprehensive background checks, and a positive work environment can contribute to this goal (Panguluri et al., 2017). Additionally, proactive communication strategies can help employees understand and better adhere to security practices that pragmatically reduce the risk of insider threats (Von Solms & Van Niekerk, 2013).

1.5 ROLE OF CULTURE IN CYBERSECURITY: SHAPING THE HUMAN FACTOR

1.5.1 The significance of cybersecurity culture

Organisational culture powerfully influences individuals' actions and attitudes within a workplace setting, including how they engage with cybersecurity matters (Panguluri et al., 2017). A cybersecurity culture, therefore, pertains to an organisation's shared norms, values, attitudes, and behaviours aimed at promoting and ensuring cybersecurity (Panguluri et al., 2017). This culture forms an integral part of an organisation's overall security posture, often playing a decisive role in determining the effectiveness of cybersecurity strategies (Panguluri et al., 2017). The role of cybersecurity culture is multifaceted. Firstly, it shapes how employees perceive and adhere to security policies (Panguluri et al., 2017). In an organisation with a robust cybersecurity culture, employees are likely to view security policies as critical tools for risk mitigation, leading to higher compliance rates.

Conversely, in organisations where such a culture is weak or absent, employees may view these policies as unnecessary burdens, leading to potential lapses in adherence (Boyce et al., 2011; Panguluri et al., 2017). Secondly, cybersecurity culture can significantly influence how seriously threats are taken within an organisation (Panguluri et al., 2017). When cybersecurity is embedded in an organisation's culture, all levels, from top management to operational staff, are likely to recognise the importance of cybersecurity and the potential ramifications of threats. Such recognition often translates into proactive actions to manage and mitigate these threats (Clark & Hakim, 2017; Panguluri et al., 2017). Finally, an organisation's cybersecurity culture can also shape how effectively incidents are handled when they occur (Panguluri et al., 2017). Organisations with a robust cybersecurity culture are more likely to have effective incident response plans and respond quickly and efficiently to security incidents. Employees in such organisations are also likely to be better prepared to identify and respond to potential security threats (Cormier & Ng, 2020; Panguluri et al., 2017).

Therefore, promoting a positive cybersecurity culture is not just about preventing threats; it also involves building resilience within the organisation and enhancing its ability to respond and recover from incidents. This includes fostering a learning environment where incidents are seen as opportunities for improvement rather than blame (Panguluri et al., 2017). Understanding and fostering a solid cybersecurity culture is integral to mitigating human factors affecting cybersecurity in critical infrastructures (Bates, 2017). Moreover, a robust cybersecurity culture forms a cornerstone for the proposed integrated cybersecurity framework, contributing to its goal of achieving systemic resilience (Cohen, 2010; Cormier & Ng, 2020). As we look to the future, organisations that successfully instil a proactive and informed cybersecurity culture will be better equipped to navigate the evolving cybersecurity landscape and effectively leverage emerging technologies such as AI.

1.5.2 Creating a cybersecurity culture in utilities

The vital role of utilities in society, combined with their increasing dependence on cyber-physical systems, makes them a high-value target for cybercriminals (Gunduz & Das, 2020). Creating a robust cybersecurity culture within utilities is paramount (Panguluri et al., 2017). This involves fostering an environment where every employee is aware of the cyber risks associated with their role, understands the importance of cybersecurity, and is committed to implementing sound cybersecurity practices (Panguluri et al., 2017). Continuous cybersecurity training and awareness programmes are vital to achieving this (Panguluri et al., 2017). Employees must be updated on the latest threats, the potential consequences of security breaches, and the protective measures they can adopt. This could include training on identifying phishing attempts, securing personal devices, and adhering to access management policies. Importantly, such training should be tailored to the unique context of utilities, considering the specific cybersecurity risks associated with their operations and technologies (Boyce et al., 2011).

Leadership commitment is another vital element in fostering a cybersecurity culture (Panguluri et al., 2017). Effective cybersecurity requires leaders to express commitment, promote integration into all processes, and hold all members accountable (Panguluri et al., 2017). Such commitment, when visible to all levels of the organisation, encourages employees to take cybersecurity seriously and strive for adherence to established policies and procedures (Cohen, 2010). Moreover, embedding cybersecurity considerations into all aspects of the organisation's operations can further enhance the cybersecurity culture (Panguluri et al., 2017). This means incorporating cybersecurity into the design and operation of cyber-physical systems, decision-making processes, and hiring and performance evaluation procedures (Cormier & Ng, 2020). It also ensures that cybersecurity policies are feasible, clearly communicated, and consistently enforced (Von Solms & Van Niekerk, 2013).

The effective creation of a cybersecurity culture in utilities helps protect these critical infrastructures from threats. It aligns with the chapter's aim to explore how human factors and organisational culture can contribute to cybersecurity (Bates, 2017). Furthermore, it provides an essential foundation for the proposed integrated cybersecurity framework by enhancing hazard management and contributing to systemic resilience (Cohen, 2010; Cormier & Ng, 2020). Utilities can build a robust defence line that complements technological security measures by considering the human factor. Ultimately, this integrated, human-centric approach represents the future of cybersecurity in utilities, one where human factors and technological advancements, like AI, work in synergy to secure critical infrastructures.

1.5.3 Shift from information security to cybersecurity

The migration from a primary focus on information security to a more comprehensive emphasis on cybersecurity signifies a strategic and conceptual evolution in response to the complexities of today's technology ecosystems (Von Solms & Van Niekerk, 2013). While the terms are often used interchangeably, a careful delineation reveals the nuanced differences and the reasons behind this paradigm shift. Traditionally, information security has been associated with protecting data in both physical and electronic forms. It emphasises the principles of confidentiality (preventing unauthorised access), integrity (ensuring accuracy and consistency over time), and availability (guaranteeing reliable access by authorised individuals) of data (Von Solms & Van Niekerk, 2013). It primarily concerns safeguarding data from threats such as data breaches, identity theft, and service disruptions (Cohen, 2010).

Cybersecurity, on the other hand, has a broader scope. It involves protecting systems, networks, and data in cyberspace (Von Solms & Van Niekerk, 2013). This includes information and the systems and processes that use, store, and transmit that information. Cybersecurity encompasses securing all interconnected digital components, such as hardware, software, and data, from cyberthreats (Cohen, 2010). This includes threats that can lead to physical damage or significant real-world consequences, especially within cyber-physical systems (Clark & Hakim, 2017; Gunduz & Das, 2020). Therefore, the shift from information security to cybersecurity reflects an understanding of the complexities and interdependencies inherent in modern cyber-physical systems (Von Solms & Van Niekerk, 2013). With this shift, organisations acknowledge that securing data alone is not enough. There is a need to protect the entire ecosystem, including the systems that process and transmit data, the networks that connect them, and the interfaces with physical systems (Cohen, 2010; Von Solms & Van Niekerk, 2013).

Given the interconnectivity of modern systems, a vulnerability in one part of the system can potentially compromise the whole, highlighting the systemic vulnerabilities inherent in these systems (Hellström, 2007). The human factors influencing cybersecurity further complicate this landscape, necessitating comprehensive strategies that address security's technical and human aspects (Bates, 2017; Boyce et al., 2011; Kovács & Nyikes, 2020). This shift also underscores the importance of developing an integrated cybersecurity framework that incorporates risk analysis, hazard management, and systemic resilience, all of which are crucial in the evolution of cybersecurity (Cormier & Ng, 2020). Furthermore, as we delve deeper into the future of cybersecurity, emerging technologies such as AI become increasingly prominent in navigating the complex landscape of cybersecurity threats.

1.5.4 Impact of culture on cybersecurity performance

The cybersecurity culture within an organisation significantly influences its ability to safeguard against, detect, and respond to cyberthreats (Panguluri et al., 2017). This culture, defined by shared values, beliefs, norms, and attitudes towards cybersecurity, plays a vital role in shaping behaviour and attitudes, consequently impacting cybersecurity performance (Panguluri et al., 2017). A robust cybersecurity culture, where the importance of cybersecurity is deeply ingrained in daily activities, has several notable benefits (Panguluri et al., 2017). Firstly, it fosters a proactive security posture where employees follow the cybersecurity policies and actively participate in enhancing them. Employees are more likely to be vigilant, comply with security protocols, identify potential threats, and report security incidents promptly (Panguluri et al., 2017). This active engagement and vigilance can help prevent potential security breaches and reduce the impact of those that occur (Boyce et al., 2011; Panguluri et al., 2017). Secondly, a robust cybersecurity culture reduces the risk associated with insider threats. With comprehensive education and awareness programmes, employees can better understand the value of the data and systems they are protecting, making them less likely to engage in harmful activities, whether intentional or unintentional (Kovács & Nyikes, 2020; Panguluri et al., 2017). Thirdly, such a culture provides resilience against social engineering attacks. With an understanding of the tactics used by attackers and the vulnerabilities they exploit, employees are better equipped to resist manipulation attempts and avoid falling for scams that could compromise the organisation's security (Boyce et al., 2011; Panguluri et al., 2017).

Conversely, a weak cybersecurity culture can significantly hamper an organisation's security performance. When employees lack understanding or awareness of cybersecurity practices and the implications of their actions, they are more likely to commit errors, ignore policies, and become targets for social engineering attacks (Bates, 2017; Panguluri et al., 2017). This can lead to a higher rate of security incidents, increased vulnerability to insider threats, and more significant potential for substantial security breaches (Bates, 2017; Panguluri et al., 2017). This cultural dimension of cybersecurity is critical to an organisation's overall security strategy. By embedding a robust cybersecurity culture, organisations can transform their employees from potential security risks into valuable assets in their cybersecurity efforts, thereby improving their cybersecurity performance (Panguluri et al., 2017).

1.6 INTEGRATED CYBERSECURITY FRAMEWORKS AND STRATEGIES: A HOLISTIC APPROACH

1.6.1 The need for integration in cybersecurity

The escalating sophistication and diversity of cyberthreats underscore the need for a more comprehensive, integrated approach to cybersecurity (Cormier & Ng, 2020). Isolated cybersecurity measures, although necessary, are no longer sufficient in today's dynamic threat landscape. This is particularly true for utilities and other critical infrastructures, which are prime targets for cyberthreat actors due to their interconnectedness and vital role in society (Cohen, 2010). An integrated cybersecurity approach represents a paradigm shift from traditional, fragmented security measures towards a more holistic and inclusive strategy (Cormier & Ng, 2020). It acknowledges that cybersecurity is not solely a technical issue but a complex, multifaceted problem involving technological, human, and organisational elements (Cohen, 2010; Cormier & Ng, 2020). It also recognises the interdependencies and feedback loops among these elements and the need to address them in a coordinated manner (Kott & Linkov, 2019).

An integrated cybersecurity strategy ensures that defences are robust and resilient across multiple dimensions (Cormier & Ng, 2020). This includes the technical dimension, which involves system architecture, software, and hardware, and the human dimension, which encompasses the attitudes, behaviours, skills, and knowledge of individuals within the organisation (Boyce et al., 2011; Cohen, 2010). It also includes the organisational dimension, covering governance structures, policy frameworks, and the overall cybersecurity culture (Panguluri et al., 2017). Acknowledging and addressing the interconnected nature of these dimensions enables an integrated cybersecurity strategy to provide a more thorough and robust defence against cyberthreats. This is crucial for protecting data and systems integrity, confidentiality, and availability and ensuring the continued delivery of essential services by utilities and other critical infrastructures (Cohen, 2010; Cormier & Ng, 2020; Gunduz & Das, 2020).

Integrating risk analysis, hazard management, and systemic resilience into cybersecurity strategies is particularly important in this context. This approach enables organisations to identify better, prioritise, and manage vulnerabilities and threats and develop resilience to withstand and recover from cyberattacks (Bates, 2017; Cormier & Ng, 2020; Kott & Linkov, 2019).

1.6.2 Principles of integrated cybersecurity frameworks

Integrated cybersecurity frameworks are based on six fundamental principles that shape their design and implementation. These principles, outlined below, encapsulate cybersecurity's holistic, adaptable, and interconnected nature (Cormier & Ng, 2020).

1. *Interdependency Recognition:* Integrated cybersecurity frameworks acknowledge that all aspects of cybersecurity, from hardware and software to human behaviours and organisational policies, are interconnected (Cormier & Ng, 2020). A failure or vulnerability in one area can potentially impact others. This principle underlines the importance of a comprehensive approach to cybersecurity that addresses individual elements and their interrelationships (Cohen, 2010).
2. *Systemic Perspective:* These frameworks adopt a systems-based perspective, which appreciates the complexity and interconnectedness of modern cyber-physical systems (Cormier & Ng, 2020; Gunduz & Das, 2020). They focus not only on the identification and mitigation

of individual vulnerabilities but also on understanding how these vulnerabilities interact within the broader system. This includes potential cascade effects, where a compromise in one area can lead to failures or breaches in others (Hellström, 2007).

3. *Continuous Monitoring and Adaptation:* Integrated cybersecurity frameworks recognise the dynamic cyberthreat landscape, with new vulnerabilities and vectors emerging (Cormier & Ng, 2020). This principle emphasises that identifying and responding to threats must be performed continuously, highlighting the need for adaptive defence strategies to accommodate changes in the threat, technological and regulatory landscapes (Kott & Linkov, 2019).
4. *Balanced Focus on Prevention, Detection, and Response:* A well-designed framework gives equal weight to preventive measures, threat detection, and incident response (Cormier & Ng, 2020). This principle aligns with the understanding that it is impossible to eliminate all risks, hence the need for robust detection and response capabilities to manage incidents effectively when they do occur (Clark & Hakim, 2017).
5. *Resilience Building:* An essential element of cybersecurity frameworks is their inbuilt resilience (Kott & Linkov, 2019). This principle stresses the importance of defending against threats and building capabilities to withstand, recover from, and adapt to cyber incidents. This could involve backup systems, redundancy, and incident response plans (Kott & Linkov, 2019).
6. *Stakeholder Engagement:* These frameworks require the involvement of all relevant stakeholders. This principle asserts that cybersecurity is a shared responsibility that requires active participation from all organisation members, from top-level management to frontline employees (Panguluri et al., 2017).

These principles underscore the alignment of integrated cybersecurity frameworks, particularly regarding systemic resilience and the role of human factors in cybersecurity (Bates, 2017; Boyce et al., 2011; Panguluri et al., 2017).

1.6.3 Risk analyses and hazard management in cybersecurity

Cyberthreats' increasing complexity and frequency underline the importance of cybersecurity risk analyses and hazard management. These two concepts are interlinked components of an integrated cybersecurity strategy and involve a cyclical process of identifying, assessing, and mitigating potential cyberthreats (Cormier & Ng, 2020).

Risk Analyses: This is the first step in the risk management process, and it entails identifying, classifying, and evaluating potential risks (Cormier & Ng,

2020). The process begins with a thorough inventory of the organisation's information assets, including hardware, software, data, and human resources. Potential threats and vulnerabilities that could result in unauthorised access, data loss, or service disruption are first identified in this step (Gunduz & Das, 2020). The potential impact and likelihood of those identified threats are then assessed to prioritise risk mitigation (Cormier & Ng, 2020).

Hazard Management: Following risk analyses, hazard management involves formulating strategies to mitigate the identified risks, considering their severity and likelihood (Cormier & Ng, 2020). Organisations often adopt strategies that involve both technical and human-centred measures to ensure the safety and security of data. Technical measures may include encryption of sensitive information, usage of firewalls and intrusion detection systems to prevent unauthorised access, timely patching and updates to address known vulnerabilities, and implementation of backup and disaster recovery plans to ensure business continuity in the event of a breach (Cormier & Ng, 2020; Kott & Linkov, 2019). Since human factors play a significant role in cybersecurity, hazard management also includes human-centred measures (Bates, 2017). This can involve creating cybersecurity awareness programmes to educate employees about common threats such as phishing and social engineering, offering training on good cybersecurity practices, fostering a cybersecurity-conscious culture, and implementing policies to manage insider threats (Bates, 2017; Boyce et al., 2011; Panguluri et al., 2017).

Integrating risk analyses and hazard management provides a proactive approach to cybersecurity, enabling organisations to anticipate and prepare for potential threats rather than simply reacting. It aligns closely with the aim of this chapter to propose an integrated cybersecurity framework that incorporates risk analysis, hazard management, and systemic resilience, and it acknowledges the significant role of human factors in cybersecurity (Bates, 2017; Boyce et al., 2011; Panguluri et al., 2017).

1.6.4 Implementing an integrated cybersecurity strategy

In the face of escalating cyberthreats, implementing an integrated cybersecurity strategy is no longer optional but necessary for organisations, especially those responsible for critical infrastructure. This multifaceted process requires active participation and commitment across all organisational levels to successfully establish a comprehensive security posture (Panguluri et al., 2017).

Leadership Commitment: Leadership is crucial in implementing an effective cybersecurity strategy. They are responsible for prioritising cybersecurity, allocating appropriate resources, and establishing a security-conscious culture within the organisation (Panguluri et al., 2017). Leaders can demonstrate their commitment by integrating cybersecurity into business strategies, promoting transparency in security practices, and endorsing continual investments in cybersecurity initiatives (Panguluri et al., 2017).

Engaging All Employees: Cybersecurity is not just the responsibility of IT departments; it is a shared obligation across the organisation (Bates, 2017; Boyce et al., 2011). Employees can strengthen or weaken the organisation's cybersecurity posture through actions. Hence, a vital aspect of an integrated cybersecurity strategy is fostering employee awareness, understanding, and adherence to cybersecurity policies and best practices. This includes ongoing education and training programmes tailored to the roles and responsibilities of different employee groups (Bates, 2017; Panguluri et al., 2017).

Ongoing Monitoring and Adaptation: In an environment of constantly evolving cyberthreats, static cybersecurity measures are inadequate. An integrated cybersecurity strategy must continuously monitor systems, networks, and user behaviours and periodic reviews and updates of cybersecurity policies and controls (Cormier & Ng, 2020). This dynamic approach enables organisations to promptly identify and respond to emerging threats, adapt their strategies as needed, and minimise the impact of security incidents (Cormier & Ng, 2020).

In addition to the above, the strategy should incorporate risk analysis and hazard management principles, as discussed earlier, emphasising technical and human-centred measures to create a resilient defence against diverse cyberthreats (Cormier & Ng, 2020; Panguluri et al., 2017). When utilities and other critical infrastructures implement such integrated cybersecurity strategies, they not only enhance their resilience against cyberthreats but also safeguard their systems, data, and the delivery of essential services and emphasise the necessity of integrated cybersecurity frameworks incorporating risk analysis, hazard management, and systemic resilience, recognising the pivotal role of human factors (Bates, 2017; Boyce et al., 2011; Panguluri et al., 2017).

Because of the multifaceted nature of cybersecurity, this chapter proposes an Integrated Cybersecurity Framework (ICF) that incorporates all of these concepts, such as risk analysis, hazard management, and systemic resilience. The ICF is structured around six key pillars, informed by the interconnected nature of cyber-physical systems and the myriad threats they face. These pillars include risk analysis, hazard management, systemic resilience, cybersecurity culture, human factors, and future-proofing (Figure 1.3).

01	02	03	04	05
Comprehensive Understanding of Cybersecurity	Insights Into Human-Related Risks	Promotion Of Integrated Cybersecurity Frameworks	Exploration of Systemic Resilience	Harnessing the Potential of AI

Figure 1.3 Proposed Integrated Cybersecurity Framework (ICF).

The first pillar of the ICF is risk analysis, which involves a meticulous process of identifying and evaluating potential cyberthreats to an organisation (Cormier & Ng, 2020). This analysis considers the specific nature of each threat, its probability of occurrence, its potential impact, and the vulnerabilities it may exploit. Given the fast-paced evolution of cyberthreats, this process is not static but is continuously updated to remain relevant.

Hazard management is the second pillar of the ICF. Once potential risks are identified and analysed, appropriate strategies are developed and implemented to mitigate these risks (Cormier & Ng, 2020). Cormier and Ng (2020) and Panguluri et al. (2017) emphasised developing and implementing targeted strategies to mitigate these identified risks. Such strategies encompass a range of measures, including technical solutions like intrusion detection systems and encryption, as well as human-centred approaches, such as comprehensive training and awareness programmes.

Systemic resilience is the third pillar of the ICF, which Kott and Linkov (2019) described as the capacity of a system to continue functioning even when faced with disruptions. That is, the resilience of a system is determined by its ability to absorb and adapt to changes, shocks, and incidents, thus enhancing its overall resilience. The objective is to ensure that cyber-physical systems can recover from cyberattacks effectively while maintaining operational integrity.

The ICF's fourth pillar is a robust cybersecurity culture built on shared norms, values, and behaviours influencing employees' cybersecurity actions (Panguluri et al., 2017). This cultural framework is pivotal in mitigating risks and enhancing an organisation's security posture.

The sixth pillar of the ICF recognises the crucial role of human operators in cybersecurity and the need to manage risks associated with human origin. As noted by Bates (2017), Boyce et al. (2011), and Panguluri et al. (2017), this includes continuous training to address insider threats and social engineering, highlighting the complex interplay between technology and human behaviour in cybersecurity.

The sixth and final pillar of ICF pertains to future-proofing. A forward-facing perspective of the virtual threat landscape is necessary to keep pace with new and emerging security challenges. Cormier and Ng (2020) and Takahashi et al. (2010) advocate for continuously adapting cybersecurity strategies and integrating emerging technologies like artificial intelligence. This proactive stance ensures that organisations can effectively detect and respond to advanced threats, safeguarding their assets against the unpredictable landscape of cyberthreats.

The ICF provides a holistic approach to cybersecurity, integrating human and technical security aspects and emphasising the system's interdependencies. By implementing this framework, organisations, especially those in charge of critical infrastructures, can enhance their resilience against cyberthreats, protect their systems and data, and ensure the continued delivery of essential services (Cohen, 2010; Cormier & Ng, 2020). The proposed Integrated Cybersecurity Framework (ICF) shares several similarities with other established frameworks, such as the NIST Cybersecurity Framework (CSF) and the ISA/IEC 62443.

Alignment with NIST CSF: The NIST CSF revolves around five functions: identify, protect, detect, respond, and recover. These functions resonate with the pillars of the ICF. For instance, 'identity' aligns with our emphasis on risk analysis, 'protect' and 'detect' coincide with our focus on hazard management, and 'respond' and 'recover' map directly onto our systemic resilience pillar. Like the NIST CSF, the ICF also underscores the importance of a continuous and dynamic approach to cybersecurity, recognising that the threat landscape evolves rapidly (Cormier & Ng, 2020). However, the ICF further emphasises the role of cybersecurity culture and human factors in cybersecurity, which are not explicitly covered in the NIST CSF. This makes the ICF particularly relevant for organisations that operate in complex, high-stakes environments where human error can have significant consequences (Bates, 2017; Boyce et al., 2011; Panguluri et al., 2017).

Alignment with ISA/IEC 62443: ISA/IEC 62443 is a series of standards developed explicitly for Industrial Automation and Control Systems (IACS). It provides detailed guidance on technical and system-level security requirements, which align with our ICF's hazard management and systemic resilience pillars (Cormier & Ng, 2020; Kott & Linkov, 2019). The ISA/IEC 62443 also strongly emphasises risk assessment, which matches our focus on risk analysis. Additionally, its emphasis on security by design and continuous improvement resonates with our recommendation for continuous adaptation and future-proofing (Cormier & Ng, 2020; Takahashi et al., 2010). Like the NIST CSF, the ICF extends beyond ISA/IEC 62443 by underlining the importance of cybersecurity culture and human factors in cybersecurity, reinforcing the significance of these elements in maintaining a robust cybersecurity posture (Bates, 2017; Boyce et al., 2011; Panguluri et al., 2017).

In closing, the ICF aligns well with existing industry frameworks while adding a more explicit focus on human factors and organisational culture. It provides a comprehensive and holistic approach to cybersecurity that reflects the complexity and interdependencies of modern cyber-physical systems.

1.7 SYSTEMIC RESILIENCE IN CYBERSECURITY: BOUNCING BACK FROM ATTACKS

1.7.1 Systemic resilience: an overview

Systemic resilience represents the robustness and adaptability of a system in the face of potential disruptions, allowing it to sustain its core operations under adverse circumstances (Kott & Linkov, 2019). Rooted in systems theory, it accentuates the interconnectedness and interdependence of a system's components. The theory advocates that a system's resilience cannot be simply deduced from the individual resilience of its components but necessitates an understanding of their collective interactions and dynamics (Kott & Linkov, 2019). In cybersecurity, systemic resilience goes beyond traditional measures of defence and protection. Instead, it emphasises the ability to endure cyberthreats, recover from security incidents, and evolve in response to changes in the threat landscape (Kott & Linkov, 2019). Due to their inherent complexity and interconnectedness, cyber-physical systems are particularly susceptible to cascading failures stemming from cyberthreats. These threats could compromise one system component and propagate across the system due to their interlinked nature, thereby magnifying the impact of the initial threat (Kott & Linkov, 2019).

Underlining the significance of systemic resilience in cybersecurity, it becomes critical for organisations, particularly those in charge of critical infrastructures, to devise strategies that enable their systems to withstand, adapt to, and recover from cyberthreats. These strategies may encompass various aspects, including strengthening security measures, establishing robust incident response plans, and promoting a security-conscious culture within the organisation. They should also involve building capacities to absorb shocks, learn from incidents, and reconfigure system components while maintaining essential functions (Kott & Linkov, 2019). By incorporating systemic resilience into their cybersecurity frameworks, organisations can ensure their capabilities to 'bounce back' from cyberattacks, mitigating the risk of catastrophic failures and preserving the continuity of their operations.

1.7.2 Role of systemic resilience in cybersecurity

In the context of cybersecurity, systemic resilience encompasses the capacity of an organisation or infrastructure to anticipate, withstand, recover from, and adapt to cyberthreats (Kott & Linkov, 2019). It emphasises preventing attacks and ensuring rapid recovery and learning during a breach (Kott & Linkov, 2019). This is critical given that no cybersecurity measures can guarantee absolute protection against all potential threats (Cohen, 2010; Kott & Linkov, 2019). Systemic resilience in cybersecurity is predicated on the understanding that no single defence mechanism can be completely foolproof. As threat actors evolve their techniques and exploit novel vulnerabilities, some attacks will inevitably succeed (Cohen, 2010; Kott & Linkov, 2019). In this scenario, resilience becomes the critical distinguishing factor between organisations that can weather a cyber storm and those that cannot.

Anticipation in this context refers to the ability to foresee and prepare for potential cyberthreats based on a comprehensive understanding of the threat landscape and an organisation's unique vulnerabilities (Gunduz &

Das, 2020). Anticipation strategies could involve ongoing threat intelligence and predictive analytics that help identify emerging threats and vulnerabilities, thus allowing proactive steps to be taken to secure the system (Gunduz & Das, 2020). Withstanding attacks, the second component of resilience, involves deploying robust defence mechanisms to prevent, deter, and disrupt attacks (Cohen, 2010). This could include the use of firewalls, intrusion detection systems, robust authentication protocols, and other technical controls (Cormier & Ng, 2020). Equally important are human-oriented strategies, such as practical training programmes that help employees recognise and respond to threats (Bates, 2017; Boyce et al., 2011; Panguluri et al., 2017). Recovery after a cyberattack involves quickly restoring normal operations and minimising the breach's impact (Kott & Linkov, 2019). This can be facilitated by incident response plans that outline precise procedures for identifying, containing, and rectifying a breach. Data backups and disaster recovery plans also play a crucial role in this stage, enabling rapid restoration of lost or compromised data (Cormier & Ng, 2020; Kott & Linkov, 2019).

Finally, adaptation involves learning from past attacks to improve future resilience. After a breach, it is essential to conduct a post-mortem analysis to understand what went wrong and how to prevent similar incidents in the future (Kott & Linkov, 2019). This may involve updating technical controls, revising policies and procedures, or enhancing training programmes (Bates, 2017; Boyce et al., 2011; Panguluri et al., 2017). The role of systemic resilience in cybersecurity thus encompasses a wide range of strategies, all aimed at enhancing the ability of an organisation to survive and thrive in the face of cyberthreats (Cohen, 2010; Kott & Linkov, 2019). By embedding resilience into their cybersecurity strategy, organisations can not only mitigate the risks of cyberthreats but also turn these challenges into opportunities for improvement and growth (Kott & Linkov, 2019; Von Solms & Van Niekerk, 2013).

1.7.3 Building systemic resilience in critical infrastructures

Building systemic resilience in critical infrastructures is not an easy task. Still, it is vital, given the critical services these infrastructures provide and the potentially devastating consequences of cyber breaches (Clark & Hakim, 2017; Kott & Linkov, 2019). This process typically involves several key steps, each of which necessitates meticulous planning, execution, and ongoing management.

Understanding the System: This is the first step in building systemic resilience. A comprehensive understanding of the system's structure, functions, and interdependencies is required to assess its vulnerability to cyberthreats (Clark & Hakim, 2017; Kott & Linkov, 2019). This involves mapping out the various components of the system, identifying the interconnections between them, and understanding how a disruption in one part could affect the whole (Clark & Hakim, 2017). This systems-thinking approach provides a comprehensive view of potential vulnerabilities and risks, enabling more effective mitigation strategies (Clark & Hakim, 2017; Kott & Linkov, 2019).

Identifying Threats and Vulnerabilities: Once the system's structure and functions are understood, the next step is identifying potential threats and vulnerabilities (Cormier & Ng, 2020). This involves analysing the cyberthreats that could target the infrastructure and examining the system's weak points (Cormier & Ng, 2020; Kott & Linkov, 2019). Risk assessment techniques, such as threat modelling and vulnerability scanning, can be used (Cormier & Ng, 2020).

Implementing Resilience Measures: Having identified potential threats and vulnerabilities, the next step is implementing measures to enhance resilience. This could involve technical defences like firewalls, intrusion detection systems, or encrypted communications (Cormier & Ng, 2020; Kott & Linkov, 2019). Equally important are human-centred measures, such as cybersecurity awareness training, transparent and enforceable policies, and a positive cybersecurity culture (Bates, 2017; Boyce et al., 2011; Kott & Linkov, 2019). The goal is to create an integrated defence that protects the system from multiple angles, recognising that no single measure is sufficient (Cormier & Ng, 2020; Kott & Linkov, 2019).

Continuous Monitoring and Adaptation: Finally, given the rapidly changing nature of the cyberthreat landscape, continuous monitoring and adaptation are crucial. This involves regularly reviewing and updating the system's defences in response to new threats or changes in the system itself (Kott & Linkov, 2019). Incident response plans and disaster recovery plans should be tested and refined regularly to ensure they remain effective (Cormier & Ng, 2020; Kott & Linkov, 2019). Emerging technologies like AI and machine learning (ML) can also enhance threat detection and response capabilities (Takahashi et al., 2010).

In sum, building systemic resilience in critical infrastructures requires a holistic approach that addresses both human and technical factors and evolves in response to changes in the threat landscape and the system itself (Kott & Linkov, 2019). By doing so, these infrastructures can better protect the essential services they provide against the increasing threat of cyberattacks (Cohen, 2010; Boyce et al., 2011; Cohen, 2010; Cormier & Ng, 2020).

1.8 THE FUTURE OF CYBERSECURITY AND AI'S ROLE: THE NEXT FRONTIER

1.8.1 Emerging trends in cybersecurity

As we look towards the future, cybersecurity is set to evolve in several vital ways. Novel technologies, methodologies, and threats continually emerge, prompting shifts in cybersecurity practices and strategies (Von Solms & Van

Niekerk, 2013). This section will highlight some of the most salient emerging trends in cybersecurity, such as the growing reliance on AI and ML, the increasing emphasis on proactivity and resilience, and the expanding recognition of the human factor in cybersecurity (Boyce et al., 2011; Boyce et al., 2011; Von Solms & Van Niekerk, 2013, pp. 6, 12).

AI and ML are increasingly instrumental in cybersecurity (Von Solms & Van Niekerk, 2013). These technologies can enhance threat detection, automate incident response, and predict potential vulnerabilities (Von Solms & Van Niekerk, 2013). For instance, ML algorithms can analyse large volumes of data to detect anomalies or unusual patterns that might indicate a cyberattack (Von Solms & Van Niekerk, 2013). AI can also simulate cyberattacks and defences, enabling organisations to proactively test and improve security measures (Von Solms & Van Niekerk, 2013). However, as these technologies are adopted, they also introduce new vulnerabilities and attack vectors that must be managed (Von Solms & Van Niekerk, 2013).

In addition to defending against attacks, there is a growing emphasis on anticipating and preparing for them (Kott & Linkov, 2019). This involves implementing proactive measures, such as threat hunting and penetration testing, and building systemic resilience to bounce back from attacks (Kott & Linkov, 2019). Cybersecurity is increasingly seen as an ongoing process of managing risk rather than a static goal of preventing all attacks (Kott & Linkov, 2019).

The role of human factors in cybersecurity will increasingly need to be recognised (Boyce et al., 2011). As systems become more complex and attackers become more sophisticated, the human element is often the weakest link in cybersecurity (Bates, 2017; Boyce et al., 2011). Recognising this, there is an expanding focus on cybersecurity culture, user education, and human-centred design in cybersecurity practices (Boyce et al., 2011; Panguluri et al., 2017).

These trends underline the need for an integrated approach to cybersecurity, which recognises the interconnectedness of various aspects of cybersecurity and balances technical measures with human-centred strategies (Boyce et al., 2011; Cormier & Ng, 2020; Kott & Linkov, 2019). This will be crucial for navigating the future's complex and evolving cybersecurity landscape (Von Solms & Van Niekerk, 2013).

1.8.2 AI and cybersecurity: a new paradigm

AI and ML technologies have dramatically reshaped the cybersecurity landscape, opening new opportunities for threat detection, prevention, and response. As Takahashi et al. (2010) articulated, these technologies herald a new era in cybersecurity, marked by increased automation, adaptability, and sophistication.

One of the primary applications of AI in cybersecurity is threat detection and prediction. Traditional cybersecurity tools rely on predefined rules and signatures to identify threats. In contrast, AI can learn from patterns and anomalies in data to detect new and unknown threats (Takahashi et al., 2010). ML algorithms can analyse large volumes of network traffic or system activity data to identify suspicious patterns that might signify an attack (Takahashi et al., 2010). Furthermore, these algorithms can be trained on historical data to predict potential attacks or identify likely targets, enabling proactive threat mitigation (Takahashi et al., 2010).

AI can also automate repetitive and time-consuming cybersecurity tasks like log analysis, alert triage, or incident response. This improves efficiency and frees human resources for more complex tasks (Takahashi et al., 2010). AI can also help orchestrate responses to detected threats, automating actions such as isolating affected systems or blocking malicious IPs (Takahashi et al., 2010).

AI's learning capabilities allow cybersecurity defences to adapt and evolve. This is crucial given the rapidly changing nature of cyberthreats. By learning from each incident, AI can continually improve its detection capabilities and response strategies (Takahashi et al., 2010). However, while AI holds tremendous promise for cybersecurity, it also introduces new challenges (Takahashi et al., 2010). For instance, attackers can use it to launch more sophisticated attacks or evade detection.

Moreover, as AI models become more complex, they can become 'black boxes', making it difficult to understand why they make certain decisions (Takahashi et al., 2010). Lastly, AI systems can be targets of attacks, such as adversarial attacks that seek to deceive AI models (Takahashi et al., 2010). Therefore, as we embrace the potential of AI in cybersecurity, it is also essential to address these challenges and ensure that AI is used responsibly and effectively (Takahashi et al., 2010). This includes securing AI systems, making AI decisions transparent and understandable, and considering the ethical implications of AI use (Takahashi et al., 2010). As we move forward into the AI-driven future of cybersecurity, these considerations will be crucial to harnessing the full potential of AI while mitigating its risks (Takahashi et al., 2010).

1.8.3 The future of AI in mitigating human-related risks

AI stands poised to revolutionise how we manage human-related cybersecurity risks. Its ability to learn from vast volumes of data, predict patterns, and automate tasks can significantly alleviate the vulnerabilities inherent in human behaviours and decisions (Bates, 2017; Boyce et al., 2011; Takahashi et al., 2010). This potential, while yet to be fully realised, points towards a future where AI plays an integral role in bolstering human defences in the cybersecurity domain.

Mitigating Human Errors with AI: Human errors significantly contribute to security breaches. These errors often stem from cognitive overload, inadequate training, or the misconfiguration of security settings (Bates, 2017). AI can help mitigate these risks by automating specific tasks and reducing the chances of human error (Takahashi et al., 2010). For instance, AI could manage complex configurations or carry out routine checks to ensure that security settings have not been inadvertently altered (Takahashi et al., 2010).

Modelling and Predicting Human Behaviour: AI and ML algorithms are proficient at analysing large datasets to detect patterns and make predictions. By applying these capabilities to human behaviour, AI can help organisations understand their employees' behaviour patterns and predict risky behaviours (Boyce et al., 2011; Takahashi, 2010). These insights can be used to develop more effective security protocols, prevent social engineering attacks, and anticipate potential insider threats (Boyce et al., 2011; Kovács & Nyikes, 2020; Takahashi et al., 2010).

Fostering Cybersecurity Culture: AI can also be crucial in fostering a solid cybersecurity culture within organisations. ML algorithms can be used to personalise cybersecurity training, tailoring it to the needs and behaviours of individual employees (Panguluri et al., 2017; Takahashi et al., 2010). This could increase the effectiveness of training programmes and improve employees' understanding and adherence to cybersecurity protocols (Panguluri et al., 2017; Takahashi et al., 2010). Furthermore, AI systems could provide real-time feedback on employees' cybersecurity behaviours, helping them to recognise and correct risky behaviours (Panguluri et al., 2017; Takahashi et al., 2010).

Envisioning the Future of AI in Cybersecurity: It is clear that while AI will augment technical defences, its most transformative potential may lie in its ability to enhance the human elements of cybersecurity. By mitigating human-related risks, modelling human behaviour, and fostering a cybersecurity culture, AI could significantly enhance the overall security posture of organisations (Panguluri et al., 2017; Takahashi et al., 2010). However, realising this potential will require careful consideration of ethical and transparency issues, ongoing research and innovation, and the commitment of organisations to leverage AI in their cybersecurity strategies (Takahashi et al., 2010).

1.9 CONCLUSIONS AND RECOMMENDATIONS: CHARTING THE PATH FORWARD

1.9.1 Summary of key findings

As we take stock of the multifaceted world of cybersecurity in critical infrastructures, several vital insights stand out. Cybersecurity is a complex domain that protects systems, networks, data, and humans. Physical and digital security are intertwined, especially in safeguarding critical infrastructures. Human factors are pivotal in cybersecurity frameworks. Systemic resilience is crucial to anticipate, withstand, recover from, and adapt to cyberthreats. Integrated cybersecurity frameworks that offer holistic defence strategies and align with industry standards are necessary. A comprehensive approach that addresses technical and human aspects, promotes systemic resilience, and incorporates advanced technologies is needed to enhance cybersecurity in critical infrastructures.

Understanding Cybersecurity's Multifaceted Nature: Cybersecurity is not monolithic. It is a complex web that includes the security of systems, networks, data, and humans (Cohen, 2010; Von Solms & Van Niekerk, 2013). The shift from focusing on information security to cybersecurity highlights the importance of understanding this intertwined nature of physical and digital security (Von Solms & Van Niekerk, 2013). Protecting critical infrastructures, like utilities, requires a comprehensive approach that addresses cybersecurity's human and technical aspects (Cohen, 2010; Panguluri et al., 2017).

The Human Factor in Cybersecurity: The role of human behaviours, emotions, and decisions in cybersecurity has been underscored throughout the discussions (Bates, 2017; Boyce et al., 2011; Clark & Hakim, 2017). Humans can be the weakest link in cybersecurity due to a lack of awareness, susceptibility to social engineering, and insider threats (Bates, 2017; Boyce et al., 2011; Clark & Hakim, 2017; Kovács & Nyikes, 2020). However, humans can also be the strongest defenders when equipped with the proper knowledge, tools, and motivation (Bates, 2017; Boyce et al., 2011; Panguluri et al., 2017). This highlights the importance of fostering a solid cybersecurity culture within organisations (Panguluri et al., 2017).

Systemic Resilience in Cybersecurity: Systemic resilience, the capacity to anticipate, withstand, recover from, and adapt to cyberthreats, has emerged as a crucial aspect of cybersecurity (Kott & Linkov, 2019). Hence, it is essential to recognise the significance of systemic resilience in reducing the likelihood of service disruptions and damages in case of a cyberattack. To build systemic resilience in critical infrastructures, one must comprehend the structure and functions of the system, identify threats and vulnerabilities, put in place measures to enhance resilience, and constantly monitor and adapt to changes (Clark & Hakim, 2017; Cormier & Ng, 2020; Kott & Linkov, 2019).

Integrated Cybersecurity Frameworks: The discussions have also emphasised the need for an integrated approach to cybersecurity. Integrated cybersecurity frameworks, like the one proposed, allow for a holistic defence strategy that recognises the interdependencies between various aspects of cybersecurity, takes a systemic perspective, and adapts to the evolving cybersecurity landscape (Cormier & Ng, 2020). These frameworks align with industry standards like NIST CSF and ISA/IEC 62443, reinforcing their relevance and applicability (Cormier & Ng, 2020; ISA, 2024).

The insights gleaned from these discussions underline the complexity of cybersecurity in critical infrastructures and the importance of adopting a multifaceted and proactive approach. They highlight the need to address both human and technical aspects of cybersecurity, foster systemic resilience, and leverage emerging technologies and methodologies like AI and ontology. Moving forward, these insights can guide the development of more robust, resilient, and adaptable cybersecurity strategies in the face of an ever-evolving threat landscape.

1.9.2 Recommendations for practice

Drawing from the insights gained from the chapter, this section provides actionable recommendations for enhancing cybersecurity in critical infrastructures. It proposes strategies to mitigate human-related risks, cultivate a cybersecurity culture, implement integrated cybersecurity frameworks, and build systemic resilience (Bates, 2017; Boyce et al., 2011; Clark & Hakim, 2017; Cohen, 2010; Cormier & Ng, 2020; Kott & Linkov, 2019; Panguluri et al., 2017).

Mitigating Human-Related Risks: Staff training and awareness programmes should be a core component of cybersecurity strategies (Bates, 2017; Boyce et al., 2011). These programmes should cover technical skills and foster awareness of cybersecurity, potential threats, and individuals' roles in maintaining security (Boyce et al., 2011; Panguluri et al., 2017). Using simulated cyberattacks or 'red teaming' exercises can help staff understand the nature of threats and how to respond (Bates, 2017; Boyce et al., 2011). Moreover, stringent access controls and monitoring systems can help to prevent and detect malicious insider activities (Boyce et al., 2011).

Cultivating a Cybersecurity Culture: A robust cybersecurity culture can significantly enhance security performance (Panguluri et al., 2017). This requires strong leadership commitment and ongoing efforts to embed security considerations into the organisation's operations (Panguluri et al., 2017). One practical approach could be integrating cybersecurity objectives into performance metrics and rewards systems (Bates, 2017; Panguluri et al., 2017). Regular communication and feedback can reinforce the importance of security and keep it at the forefront of employees' minds (Panguluri et al., 2017).

Implementing Integrated Cybersecurity Frameworks: An integrated approach to cybersecurity can provide robust protection against critical infrastructures' multifaceted threats (Cormier & Ng, 2020). Frameworks such as the NIST Cybersecurity Framework or the IEC 62443 standard provide a comprehensive guide for implementing such an approach (ISA, 2024). These frameworks emphasise a continuous process of identifying, protecting, detecting, responding to, and recovering from cyberthreats (ISA, 2024). They also underscore the importance of considering technical and human factors in cybersecurity (Cormier & Ng, 2020; ISA, 2024).

Building Systemic Resilience: Systemic resilience involves understanding the system's structure, identifying threats and vulnerabilities, implementing resilience measures, and adapting to changes and new threats (Clark & Hakim, 2017; Cormier & Ng, 2020; Kott & Linkov, 2019). Incident response plans should be implemented to ensure rapid and effective attack recovery (Kott & Linkov, 2019). Continuous monitoring and regular 'health checks' can help identify and address potential vulnerabilities before they are exploited (Cormier & Ng, 2020; Kott & Linkov, 2019).

Leveraging AI and Ontology: AI and ontology hold significant potential for enhancing cybersecurity (Gunduz & Das, 2020; Takahashi et al., 2010). AI can automate threat detection, conduct predictive analysis, and respond to threats more quickly and effectively (Takahashi et al., 2010). On the other hand, ontology can help to improve understanding and communication in cybersecurity, making it easier to identify and address vulnerabilities (Gunduz & Das, 2020). Organisations must invest in the necessary skills and infrastructure to leverage these technologies (Takahashi et al., 2010). Collaborating with technology vendors, academic institutions, or specialised cybersecurity firms may be beneficial (Takahashi et al., 2010).

In conclusion, securing critical infrastructures is complex and challenging but also essential. The recommendations in this section offer practical strategies for enhancing cybersecurity, which can be tailored to individual organisations' specific needs and contexts. It is crucial to remember that cybersecurity is a continuous effort, requiring ongoing monitoring, learning, and adaptation in the face of a constantly evolving threat landscape.

1.9.3 Avenues for future research

Future research in cybersecurity has great potential to improve our understanding and development of more effective strategies and frameworks. There is a particular need to explore the human factor in cybersecurity, including how human behaviour affects cybersecurity. This could involve examining the impact of demographic factors and organisational cultures on security practices and developing innovative methods for altering behaviours and promoting a culture of cybersecurity awareness (Bates, 2017; Boyce et al., 2011; Panguluri et al., 2017).

Another critical area of focus is creating and validating integrated cybersecurity frameworks. This might include developing or refining new frameworks and conducting comparative studies or empirical research to assess their effectiveness in various settings. It is crucial to explore how these frameworks can be implemented practically across different organisational contexts, considering various constraints (Cormier & Ng, 2020; ISA, 2024).

The concept of systemic resilience in cybersecurity also warrants further investigation. Understanding how to achieve systemic resilience across different types of organisations and infrastructures is essential. This includes studying the factors contributing to resilience and developing methods to measure and assess it, which is vital for tracking progress and identifying areas that require improvement (Kott & Linkov, 2019).

As the use of artificial intelligence (AI) in cybersecurity continues to increase, future research should aim to advance AI technologies and explore their potential applications in this domain. This could involve creating new AI algorithms or techniques specifically for cybersecurity purposes and considering the ethical and practical challenges of using AI in this field. Additionally, the role of ontology in supporting various cybersecurity aspects, such as risk assessment and incident response, presents another area for research (Gunduz & Das, 2020; Takahashi, 2010).

Lastly, conducting in-depth case studies and empirical research on cybersecurity within critical infrastructures is crucial. These studies can provide insight into real-world challenges and strategies employed in cybersecurity, offering invaluable insights that can inform both practice and policy. Furthermore, empirical research can validate and refine theoretical concepts and frameworks, thereby enhancing the academic rigour of the cybersecurity field (Hellström, 2007).

In conclusion, while much has been learned about cybersecurity, much remains to be explored. The suggested avenues offer exciting opportunities for advancing knowledge and practice in this vital field. Future research in these areas can significantly enhance the security and resilience of critical infrastructures worldwide.

Cybersecurity's growing importance and complexity in critical infrastructures cannot be understated. This chapter has illuminated various facets of cybersecurity, drawing particular attention to the human factor, the evolution from information security to cybersecurity, the role of culture in shaping cybersecurity practices, the need for integrated cybersecurity frameworks, the concept of systemic resilience, and the promising role of AI in the cybersecurity landscape. The human factor, often overlooked, is a critical component of cybersecurity. Fostering a robust cybersecurity culture within organisations is not just about implementing strict policies or employing the latest technological defences. Still, it also involves educating employees, nurturing an environment of shared responsibility, and embedding cybersecurity considerations into all aspects of the organisation's operations.

Moving from information security to cybersecurity represents a broader view of the intertwined nature of physical and digital security, highlighting the need to address human and technical security aspects. The shift reflects the complex interdependencies within cyber-physical systems and recognises the reality of the contemporary digital environment. Building systemic resilience in critical infrastructures and implementing integrated cybersecurity strategies are essential for protecting these critical systems. They require a deep understanding of the systems, their potential threats and vulnerabilities, and the continuous adaptation to evolving circumstances.

AI promises a new paradigm in cybersecurity, with its ability to automate tasks, predict attacks, and learn over time. It holds particular potential for mitigating human-related risks and contributing to developing a robust cybersecurity culture. However, despite the advancements in understanding and addressing cybersecurity challenges, the field is evolving, and new threats, methodologies, and technologies continue to emerge. As such, continual research and adaptation are vital to staying ahead of these threats. Future research is encouraged to delve deeper into the mentioned topics, aiming for a safer and more secure digital future. This chapter highlights that cybersecurity in critical infrastructures is a multifaceted and evolving challenge that requires a comprehensive, integrated, and forward-looking approach. It underscores that everyone within an organisation has a role in cybersecurity and the importance of resilience, adaptability, and continuous learning in navigating the cybersecurity landscape. As we look towards the future, the insights and discussions presented in this chapter can provide valuable guidance for practitioners and researchers in enhancing the security and resilience of our critical infrastructures.

REFERENCES

Bates, S. (2017, May). Managing human factors. In *62nd Annual Business Aviation Safety Summit, BASS 2017* (pp. 105–110). https://doi.org/10.1007/978-3-319-77492-3_11/COVER

Boyce M. W., Duma K. M., Hettinger L. J., Malone T. B., Wilson D. P., Lockett-Reynolds J. (2011). Human Performance in Cybersecurity: A Research Agenda. *Proceedings of the Human Factors and Ergonomics Society Annual Meeting*, 55(1), 1115–1119. https://doi.org/10.1177/1071181311551233

Clark, R. M., & Hakim, S. (2017). *Cyber-physical security: Protecting critical infrastructure at the state and local level*. http://dx.doi.org/10.1007/978-3-319-32824-9_1

Cohen, F. (2010). What makes critical infrastructures critical? *International Journal of Critical Infrastructure Protection*, *3*(2), 53–54. https://doi.org/10.1016/J.IJCIP.2010.06.002

Cormier, A., & Ng, C. (2020). Integrating cyber security in hazard and risk analyses. *Journal of Loss Prevention in the Process Industries*, *64*, 104044. https://doi.org/10.1016/J.JLP.2020.104044

Gunduz, M. Z., & Das, R. (2020). Cyber-security on the smart grid: Threats and potential solutions. *Computer Networks*, *169*. https://doi.org/10.1016/j.comnet.2019.107094

Hellström, T. (2007). Critical infrastructure and systemic vulnerability: Towards a planning framework. *Safety Science*, *45*(3), 415–430. https://doi.org/10.1016/J.SSCI.2006.07.007

ISA. (2024, February 7). ISA/IEC 62443 series of standards. *ISA*. Retrieved from www.isa.org/standards-and-publications/isa-standards/isa-iec-62443-series-of-standards

Kott, A., & Linkov, I. (2019). Cyber resilience of systems and networks. *Cyber Resilience of Systems and Networks*. https://doi.org/10.1007/978-3-319-77492-3

Kovács, T. A., & Nyikes, Z. (2020). Uncertainties in safety and security: Uncertainties in critical infrastructure protection and human factors. In *Safety and Security Issues in Technical Infrastructures* (pp. 383–412). http://dx.doi.org/10.4018/978-1-7998-3059-7.ch014

Panguluri, S., Nelson, T. D., & Wyman, R. P. (2017). Creating a cyber security culture for your water/waste water utility. In *Cyber-physical Security* (pp. 133–159). https://doi.org/10.1007/978-3-319-32824-9_7

Takahashi, T., Fujiwara, H., & Kadobayashi, Y. (2010). Building ontology of cyber security operational information. *ACM International Conference Proceeding Series*. https://doi.org/10.1145/1852666.1852756

Von Solms, R., & Van Niekerk, J. (2013). From information security to cyber security. *Computers & Security*, *38*, 97–102. https://doi.org/10.1016/J.COSE.2013.04.004

Chapter 2

Analysing cyber-physical attacks

The human operator challenge in mining

Osmar Couto and Ahmed Ibrahim

2.1 INTRODUCTION

Supervisory Control and Data Acquisition (SCADA) systems are embedded in most process plants in mining sites worldwide. Although mining activities do not represent a mission-critical system for human survival, they provide a base for a high economic proportion for specific countries where this sector is the leading tax income generation. Therefore, these mining sites represent significant targets for cyber-physical attacks (CPA) to destabilise and cause financial loss or even be classified as terrorist actions when they threaten a country's stability.

Real-world incidents have flagged several industries and started decades ago (Nicholson et al., 2012; Pizarro & Fuenzalida, 2021). Nicholson et al. (2012) explained that SCADA attacks were 30% external and 70% internal in a survey between 1982 and 2000, which had turned to the opposite during 2001 and 2003, where 30% were internal and 70% were external attacks; the authors of the survey attributed the increase in external attacks to the standardisation of the network standards. Coates et al. (2010) presented details of SCADA (Supervisory Control and Data Acquisition) system attacks that were not widely disseminated. These attacks included instances such as the use of a trojan firmware by the United States to trigger an explosion in a pipeline located in the Soviet Union during the Cold War. In another example, Coates et al. (2010) identified cyberattackers that could infiltrate the primary power grid in California, in 2001, and the intrusion went unnoticed for a staggering 17 days. These incidents serve as stark reminders of the potential vulnerabilities of critical infrastructure systems to cyberthreats and the potentially devastating consequences of successful attacks.

Further, in 2003, Slammer Worm turned off the safety monitoring of the Davis-Besser nuclear plant in Ohio. Worldwide known attacks were used to warn about SCADA system vulnerabilities, employing malware such as Stuxnet (2010), Duqu (2011), and Havex (2014) (Nazir et al., 2017). Moreover, the Ukrainian power industry and railway were attacked by BlackEnergy crimeware (Adepu et al., 2020).

DOI: 10.1201/9781032664859-2

A complex SCADA system is usually supervised by human operators who follow up on alarms and anomalies through a human–machine interface (HMI) and other supervisory interfaces that communicate the health of the system and monitor endpoint devices, such as actuators, limit switches, pressure valves, and other sensors.

The human operator reacts to abnormal behaviour of the mining process plant (MPP) based on theoretical training and practical knowledge of the process plant or how it should work as business as usual (BAU) behaviour. The decision to stop production is not taken lightly, as the implications are costly. Therefore, the human operator always performs the best to keep the plant running at maximum capacity.

Cyber-physical systems (CPS), defined by the interaction between humans, cyberspace, and machines, are employed in smart grids, traffic controls, e-health, and smart infrastructures (Kriebel et al., 2018). The Industrial Control System (ICS) is also part of the CPS realm, presenting different features and concerns than the Information Technology (IT) realm.

2.2 MINING PROCESS PLANTS (MPP)

MPPs vary depending on several characteristics, such as the mining type, mineral and final product required, and SCADA system. Therefore, a basic overview of these characteristics is presented to establish an environment for the MPP model proposed for the analysis in this chapter.

2.2.1 Mining

MPPs are usually designed based on the mining type where the mineral can be found, which dictates the type of processing plant: open-pit, water table, surface mining, and underground mining (Parreno, 2017; Yihdego & Drury, 2016). There are several types of minerals, as shown in Figure 2.1. Moreover, each type of mineral requires a specific process frame, such as drilling, crushing and screening, grinding, slurry, pyro, material handling, and compaction (Metso Corporation, 2018).

MPP can be complicated and designed in several stages. Figure 2.2 shows the generic concept of a process plant as a guideline to understand the basics of a process plant environment at a mining site. Table 2.1 presents the basic process concepts for an iron ore mining site. These processes are not mandatory and can be arranged in different scenarios, as required for the outcome of the production of the desired product.

Completing all these processes requires a complex process control interaction with the process plant infrastructure as temperature, chemicals, volume, high torque machines, loading and unloading machines, and handling tonnes

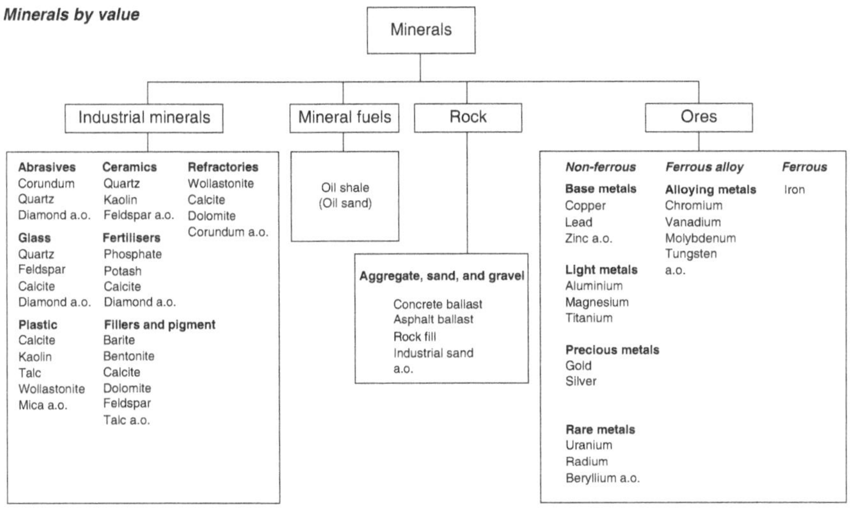

Figure 2.1 Several Mineral Types Explored in the Mining Industry (Metso Corporation, 2018).

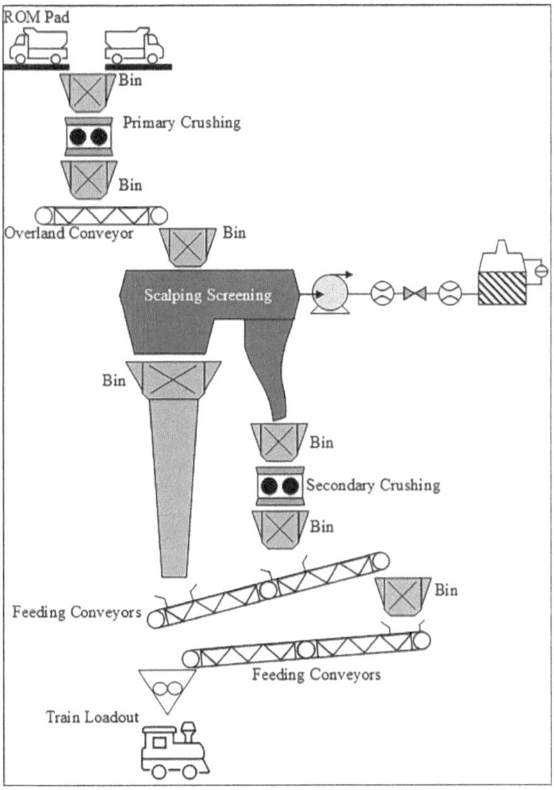

Figure 2.2 The Generic Concept of a Process Plant.

Table 2.1 Processes and Brief Descriptions

Process	*Description*
ROM pad	The crude mineral enters the process plant through the ROM Pad
Primary crushing	The mineral is crushed into the desired size for processing, and the coarse ore is collected
Screening	The required size is achieved during the secondary crushing stage, and any remaining fines can be collected
Material handling	Material handling is responsible for moving minerals across all processes, including overland and feeding conveyors, bins, train loadout, and all other equipment specific to each process or mining site
Train loadout	The train loadout system loads materials onto wagons for despatch to clients

of minerals must be precisely controlled (Table 2.1). The safety instrumented systems (SIS) run with the primary process control system and provide a fail-safe system to take over the primary system in the case of main system failure. Moreover, the fail-safe system must be up to date and tested to ensure that if the main process controls fail, the SIS will take over and guarantee that all systems fall back to a safe position and put no lives at risk.

The main components of a mine processing plant are illustrated in Figure 2.2, beginning with crude ore dropped into a crushing bin and ending with the final product delivered to a train loadout. The MPP begins by crushing and screening the ore, moving it to the scalp and screening it before the final products are shipped to clients or buyers by train. Not depicted in the diagram are additional utility services that support the processing plants operations and include power, water, and communications. Operators can control the processes and take action if critical plant systems, such as the SCADA network, fire alarm monitoring system, or, in some locations, a tornado system raise the alarm.

2.2.2 Overview of SCADA systems in MPP

SCADA systems generally show the process plant stages broken down into pages. The SCADA function informs the operator of how the plant is working and its status. Generally, one or more SCADA system pages capture each MPP process or stage. This page scheme is employed to facilitate the operator to follow the stage's functionality or to track better the information presented on the page, which could come from sensors, instruments, manual inputs, and data recorded for comparison with previous stages or the operator's shifts.

The generic plant diagram from Figure 2.2 would be shown on five or more pages, grouping the main instruments and systems. SCADA pages can vary from a simple follow-up on crushing and transporting ores in conveyors to a complex process of treating a mineral with mills, valves, instruments, pumps, pressure controls, and other devices that can add variables to the processes and vulnerable points to CPA. An alarm on the SCADA system page can require several levels of action from the operator. Table 2.2 presents the action levels the human operator is required to follow according to the alarm level. Figure 2.3 shows the concept of a SCADA alarm page and its alarm levels.

Table 2.2 The Action Level Required by the Human Operator According to the Alarm Level

Alarm Level	*Human Operator Action*
Minimal (green)	The operator acknowledges the alarm and leaves the plant running
Medium (orange)	The operator acknowledges the alarm and stops part of the plant affected by the anomaly that triggered the alarm, trying to isolate the attack and avoid letting the attacker reach the entire plant
Maximum (red)	The operator shut down the entire plant until the issue that triggered the alarm was verified and fixed

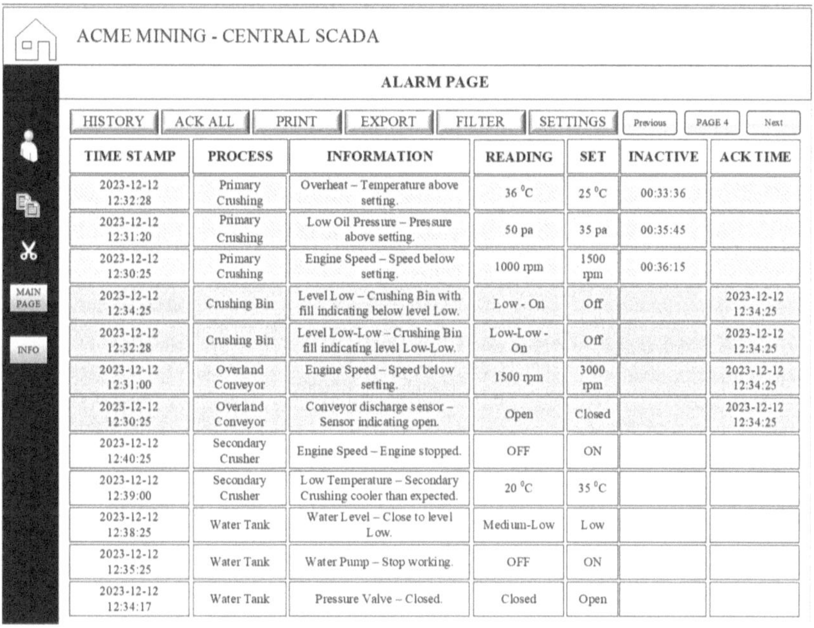

Figure 2.3 SCADA Alarm Page Visualisation.

Note: This figure shows a SCADA alarm page concept. The shade of the alarm row can identify different threat levels to the MPP operation: no shade, less concerning, already treated, going up to heavily shaded as a more threatening issue, or a lightly shaded alarm that was not acknowledged and escalated to a deep shade.

2.3 CYBER-PHYSICAL SYSTEMS AND ATTACKS

2.3.1 Cyber-physical attacks (CPA)

Cyber-physical systems (CPS) are vulnerable to both cyberattacks and physical attacks. When cyberattacks result in physical damage, they are classified as CPA (Yampolskiy et al., 2012). CPA can harm the CPS infrastructure in a process plant and lead to economic losses and fatalities in rare cases (Li et al., 2016). Understanding CPS vulnerabilities requires identification and measurement of situational awareness of virtual cyberthreats and local physical attacks.

Focusing on SCADA systems, Nicholson et al. (2012) analysed the role of SCADA systems in the cyber realm and its vulnerabilities. They broke down all components of possible cyber warfare manoeuvres that could target private companies, government, or the military. They explained that adversaries in OT are the same as in the IT environment, and the energy sector was more affected than others. Moreover, disgruntled employees were among the top adversaries responsible for cyberattacks. The consequences of cyberattacks on SCADA can be severe, causing loss of lives, financial loss of revenue, share price reduction, or even brand damage (Nicholson et al., 2012).

Nazir et al. (2017) warned that several vulnerabilities exist in the SCADA systems, such as the ones described in Table 2.3. Further, the poor administration of SCADA systems by their owners and the vendor's responsibility to fix their products' system vulnerabilities could be added to the list of vulnerabilities, even after they retire or stop the equipment sale (Nicholson et al., 2012). Do et al. (2017) and Nicholson et al. (2012) have identified various types of SCADA system threats that can be classified based either on the attack techniques used or specific system targets (Table 2.3).

Table 2.3 CPS Vulnerabilities and Threats (Do et al., 2017; Nazir et al., 2017)

Vulnerabilities	*Threats by Attacking a Specific Target*	*Threats by Attack Technique*
• Generic operating systems • Legacy systems with long operational life • Multiple points of entry and failure • Communications protocols • Real-time and complex interactions • Conflicting priorities (patching) • Social engineering and insider attacks • Backdoors • Integral protection	• Cyberattacks on the control centre where the SCADA servers and back-end infrastructure are hosted • Cyberattacks on a communication network – all interconnections between the different layers of the CPS • Physical attacks on the technical process – physically damaging the infrastructure or its components	• Malware • Denial-of-service (DoS) • Network attacks • Main-in-the-middle attacks • False sequential logic (FSL) attacks • False data injection (FDI) • Integrity attacks

The literature presents several examples of these attacks, such as the FSL proposed by Li et al. (2016), where they explained that CPS work is based on a feedback response built on the sequence of commands and sensor readings. Therefore, an incorrect sequence of legitimate commands would disturb the sequence in which processes occur, which would cause damage to or interruption of the CPS.

2.4 HUMANS AND HUMAN OPERATORS IN MINING

2.4.1 Human operations in mining

Mining is a high-risk industry with high accident rates. A systematic review of research trends in mining accidents identified 57 studies related to mining accidents from 2015 to 2019 (Noraishah Ismail et al., 2021). Four main themes revealed were as follows:

1. The leading causes of mining accidents (46%).
2. Prevention of mining accidents (20%).
3. Challenges (17%).
4. Impacts (17%) of post-mining accidents.

The primary cause of mining accidents, accounting for more than half of the studies reviewed by Noraishah Ismail et al. (2021), was mechanical failure, followed by poor safety management and leadership issues. To prevent such accidents, recommended strategies include using software applications and models, training programmes, improved regulations, and supervision. Mine owners face challenges in ensuring safe work environments and dealing with the consequences of accidents. Most of the reviewed articles, totalling over 60%, focused on coal mining accidents in China. Other countries represented in the studies included Brazil, South Africa, Spain, Iran, the United States, India, and Turkey. Mining accidents harm human health, the environment, societies, governments, and economies, with significant environmental disasters such as Brazil's 2015 Fundão tailings dam collapse (Noraishah Ismail et al., 2021).

The psychological pressure on Operators is anticipated as they oversee various systems while people work around and with them. This pressure arises from the potential threat to human lives from cyberattacks that could disrupt one or more of these systems, leading to mechanical failures in numerous equipment used in mining processes.

2.4.2 Human operator

The plant operator or control room operator, further referred to as the Operator, supervises the entire plant process or partial processes via the SCADA system. To become an operator, they must undergo technical training regarding the equipment and machinery environment in which they work. However, there is no training in cybersecurity or how to cope with the pressure of an imminent CPA. In addition, there is a mining social environment to be considered as isolation, as many mining sites are in remote areas, which can bring mental health issues.

The Operator is trained in the processes designed and implemented in the mining site plant and supports activities to keep the workplace environment safe. Some examples of training available in Australia are as follows:

- Control Room Operator Course by Coal Services (Services, 2018)
- Certificate II in Process Plant Operations (PMA20116) by TAFE (Australia, 2019)
- Certificate III in Underground Metalliferous Mining by TAFE (Training, 2018)

Table 2.4 lists the units in PMA20116 related to IT/OT, which do not include any cybersecurity knowledge or skills required by the MPP. Most training programmes are designed to train the Operator to be competent in operating SCADA systems in a real-world environment. RMIT's (2019) SCADA systems course and Nexus' (2019) workshop are good examples.

The Operator monitors the SCADA HMI page in the control room. If an alarm is triggered, the HMI alarm page of the SCADA system alerts the operator that the process plant is in an abnormal operating state. The primary action of the operator is to identify the cause of the alarm. The operator's decision on what to do next depends on two sources of information: their Subject Matter Expertise (SME) in the process plant's operations and the information displayed on the HMI screen from the SCADA system. Operators typically trust their expertise, which is a combination of their job experience and training. However, operators may not always trust the information a SCADA system provides. As an SME in the process plant's operation, the operator is not an expert in CBPS and, therefore, lacks the necessary knowledge to assess CPA, which is beyond the scope of their job description.

Table 2.4 Extract of the PMA20116 Process Plant Operations Certificate II Syllabus Related to IT/OT (Australia, 2019)

Code	*Title*
MSMOPS212	Use organisation computers or data systems
MSS402060A	Use planning software systems in operations
PMAOPS236	Monitor continuous process plant
PMAOPS216	Operate local control system
PMAOPS280	Interpret process plant schematics

2.5 HUMAN PSYCHOLOGICAL CHALLENGES IN MINING

2.5.1 Impact of cyberattacks on mental health

Bada and Nurse (2020) identified in their research that cyberattacks that increase threats can have significant social and psychological impacts. The critical issues examined included risk perception, protection motivation, culture of fear, and attacker/target characteristics. Real-world examples such as the WannaCry ransomware and Lloyds Bank cyberattacks were analysed to demonstrate the concepts.

Their principal findings, presented in Table 2.5, were associated with victims developing negative emotions like anxiety, trauma, and helplessness based on perceived control, as their response efficacy and self-efficacy shape threat perceptions and severity. Susceptibility to attacks was less than desirable or acceptable in their minds. Moreover, post-attack government communication and the attacker's identity can influence public response and further damage the victim's confidence.

When practitioners in different professions and industries are responsible for the confidentiality of their clients or patients, they are hit by an uncomfortable feeling of responsibility for the data and the risk of losing them. For example, the impact of cyberattacks on mental health in the digital mental health industry was investigated by Inkster et al. (2023).

Inkster et al. (2023) ponder the damage caused by the weight of such responsibility, as their research suggests that practitioners need to be aware

Table 2.5 Findings from Gross et al. (2016) and Bada and Nurse (2020)

Impact	*Title*	*Reference*
Social	Disruption to daily lives, widespread anxiety, loss of confidence in technology	Bada and Nurse (2020)
	Isolation or depression after financial losses	
	Erosion of trust in technology after attacks cause service disruptions	
	Demands for greater protection and regulation from government post-attack	Gross et al. (2016)
Psychological	Negative emotions like frustration, anger, anxiety, and helplessness	Bada and Nurse (2020)
	Trauma, depression, grief, shame, and sense of violation after victimisation	
	Worry about potential life endangerment from critical infrastructure attacks	
	Loss of perceived control and learned helplessness	
	Paranoia, avoidance behaviours, and lack of confidence after cyberattacks	

of the potential risks of cyberattacks and take appropriate measures to protect client confidentiality, not only by storing them but also during virtual assessments. Furthermore, they transpose the same assessment to cybersecurity professionals, who also need to be aware of the potential risks of cyberattacks and take appropriate measures to protect their mental health.

The psychological impact of a cyberattack can be tremendous, and the effects can be felt for weeks and months. Cyberattack victims may feel embarrassed, ashamed, and worried about what others think if they discover the details. Workplace security mistakes can lead to employment loss, which can devastate one's mental and physical well-being. Avoidance is also common; if a person feels uncomfortable thinking about a cyberattack, they might use discomfort as an excuse to avoid improving their cybersecurity. Ignoring their feelings can prevent them from processing what happened and doing what they must to ensure that it does not happen again (Steel, 2022).

2.5.2 Human operator mental health

Pizarro and Fuenzalida (2021) presented findings that mining operations can significantly impact their workforce's mental health and safety. After reviewing the relevant literature, it appears that managing these issues requires addressing four key themes: psychological and personal factors, psychosocial and health-related factors, physical well-being, and organisational factors. The research found that mine workers are at risk of various mental health issues, including depression, anxiety, stress, and PTSD.

The survey addressed literature from 2008 to 2019 in several countries such as Australia, Chile, China, and others that are not less critical while varying the mining activity from coal, copper, gold, diamond, and others that are not specified. Other variables were the research method applied, instruments, and sample qualification by gender and age (Pizarro & Fuenzalida, 2021).

This research suggests that mine workers are at risk of various mental health issues, and it is essential to identify and manage these risks to create a more mentally healthy workplace. To ensure their well-being, enterprises must address various factors, such as hazardous environments, work organisation, interpersonal relations, psychosocial risks, well-being, substance abuse, personality traits, and physical health.

2.6 AUTONOMOUS CYBER-PHYSICAL SECURITY (CBPS)

Researchers have proposed different approaches to autonomous CBPS to replace the cybersecurity specialist in detecting CPA. The following research articles have presented the essential characteristics contributing to this chapter's discussion.

Layal et al. (2016) proposed an autonomous SCADA system using discrete components and software using a wireless network to connect the sensors. The Autonomous SCADA System research is a straightforward example of how setting up and controlling thresholds can be used for process control supervision and automation of a process plant, as monitored abnormal behaviour can be used to alarm or trigger a response. The Recommender and the Anomaly Behaviour Analysis (ABA) showed one way to set up the interrelation between the assistant and the SCADA system.

Sayan et al. (2017) proposed a cybersecurity assistant based on machine learning. The assistant was trained to identify normal and abnormal behaviours using cyber situation analysis and recognition features by predicting attacks and employing cyberattack trend and impact analysis modules. This solution employed IDS detection based on network traffic to analyse abnormal behaviour, which could not identify FSL and FDI attacks.

The Behavioural Observation for Critical Infrastructure Security Support (BOCISS) proposed by Hurst et al. (2013) is an independent agent from the SCADA system that collects data from the SCADA system through network access to the plant equipment. When abnormal behaviour is identified, a data manager collects and distributes data to BOCISS, classifying, storing, preparing, and presenting the alarm to the operator.

A self-protecting SCADA system model introduced by Chen et al. (2016), which uses function code scanning and malicious parameter injection attacks to validate self-protecting features, was proposed. A dataset created in real time from the system behaviour is sent to the intrusion estimation module (IEM), which predicts the expected behaviour and then compares it to the system's actual behaviour. When deviations from the predicted value are found, warnings are sent to the multi-criteria controller (MAC), which triggers a response to a possible attack.

A novel intrusion tolerance scheme for a networked control system was proposed by Huang et al. (2015) as a case study in the Ocarina Architecture Analysis and Design Language (AADL) environment. Three types of injected attacks were claimed to be defended by the proposed scheme.

A System-Aware Cybersecurity Architecture proposed by Jones and Horowitz (2012) combines techniques, such as multiple vendors, to create diversity, configuration hopping, data consistency checking, and tactical forensics. The system works through a real-time collection of information to assess the integrity of the plant and then analyses the information collected and acts with controls to protect the system, if required. A case study was proposed using a nuclear plant, where an examination was presented considering the pre-attack, trans-attack, and post-attack phases.

The literature discussed above explores the evolution of cybersecurity solutions, from simple microcontrollers to fully developed machine learning operators. The gaps in these six types of research highlight the need for a more comprehensive solution, as the operator is a crucial component of the CPS. They are responsible for detecting attacks on the CPS and taking the necessary action to mitigate any potential attacks and protect the CPS. Thus, investigating and mitigating attacks are procedures a human operator trained in troubleshooting procedures for SCADA systems would perform based on their training in response to a SCADA alarm.

2.7 AUTONOMOUS OPERATOR FOR CYBER-PHYSICAL SYSTEMS

The role of the human operator is a crucial but often neglected aspect of the intricate field of safeguarding mining process plants from cyber-physical attacks. Although the existing literature has mainly concentrated on automated methods that employ sensor data, simulated situations, and network analysis, there is a lack of comprehension and integration of the human element.

This section aims to explore the diverse obstacles faced by human operators in the context of mining operations and to recognise their indispensable role in the face of a continuously evolving threat environment. It delves into the relationship between psychological stress, operational risks, and cyberthreats and proposes incorporating autonomous agents to bridge the gap between cybersecurity knowledge and the complex workings of mining process plants.

The focus of protection measures against cyber-physical attacks for mining process plants has mainly been on automated methods that depend on data from various sensors, simulations, and network traffic analysis. Although these strategies promise to strengthen system security, there is a noticeable gap in the field: the limited incorporation of the human operator. As technology progresses, it is vital to address the human element, which includes cognitive complexities, vulnerabilities, and crucial roles that have been relatively disregarded in current research.

Human Operators in an MPP must contend with the tangible dangers associated with their job and deal with the dual pressures of psychological stress and the constant threat of cyberattacks. The convergence of these pressures creates a complex environment where the human operator is not only expected to navigate the intricate complexities of mining operations but also to bear the responsibility of protecting against potential cyber breaches, which adds to the existing stressors.

Improving the relationship between cybersecurity and the practical requirements of mining operations calls for a change in how we think about the situation. One emerging solution is using autonomous agents, in this case, an Autonomous Operator (AO), that can bridge the gap between cybersecurity and the real-time demands of mining process plants. The AO has the potential to enhance the abilities of the human operator by serving as an intermediary that combines extensive cybersecurity knowledge with the intricacies of mining operations, thereby protecting against physical cyberattacks.

The Autonomous Operator (AO) would react to the alarms presented by the SCADA system, reproducing the procedure that would be followed by a human operator trained in troubleshooting procedures in SCADA systems to decipher the cause of the alarm. If the cause of the alarm is deemed a malicious attack, the AO will collect information from all layers of the MPP, normalising the information in a way that the AO can process. Once the AO can ingest the information, the classification would determine if CPA caused the abnormal behaviour. When CPA is identified, the AO communicates to the human operator via the SCADA HMI that the cause of the abnormal behaviour is CPA.

The AO may lift the burden from the Operator to identify or be concerned with CPA. With that purpose and its effects, the mental health issues caused by the continued concern of the operation's disruption and blame if successful CPA occurs can be reduced and, with time, eliminated, giving back to the Operator their confidence. Furthermore, the AO can work as a machine companion to learn from and double-check actions over time. The AO does not mean to be a cybersecurity specialist; it aims to augment the Operator's skills. However, its assessment is second to none because it is a machine learning algorithm prioritising access to specific information.

The AO needs to trust the information sent to it to create independent connections between the AO and critical information points in the proposed process plant model. Even with separate connections, information gathered by the AO can be compromised. If an attacker identifies these connections, an attempt to interfere with or alter the data collected by the AO could occur, which could mislead the AO's classification.

2.7.1 A Theoretical CPS model for an autonomous operator

Figure 2.4 presents a theoretical CPS model high-level block diagram of a typical MPP block diagram to represent the CPS on the MPP for implementing AO in an MPP. The AO will be independently connected, through a firewall, to the critical points of the MPP to collect data without relying on the SCADA network, which could be compromised during CPA. The direct access to critical points, such as PLCs, HMI stations, legacy and SCADA servers, allows the AO to compare the data in different stages of the processes. Hence, the AO can identify where the data or the process was altered or attacked. The data flow typically goes from the valve/actuator to the PLC, is visualised or handled by the SCADA system, and is recorded to the historian server.

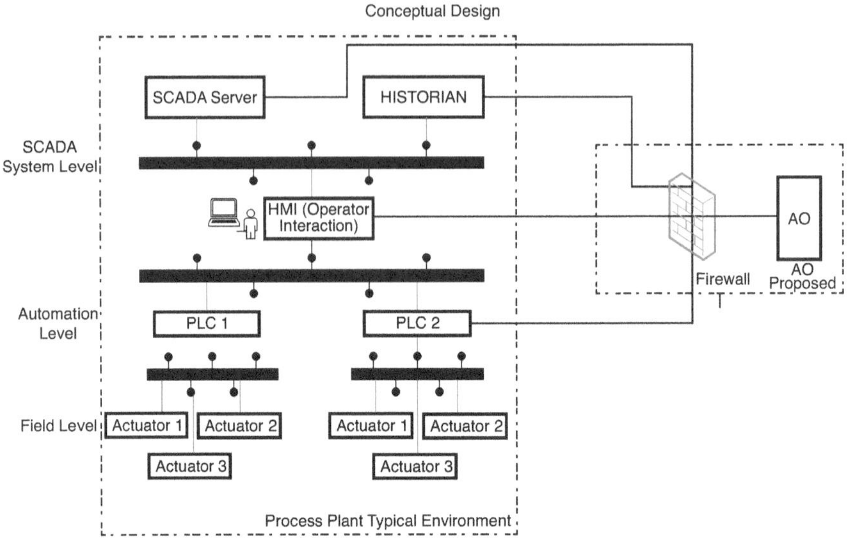

Figure 2.4 The Conceptual Design of the CPS Model Proposed.

Two significant advantages of an AO in an MPP are as follows:

1. Troubleshooting and conducting investigations
2. Truly autonomous

The SCADA system operator knows how to troubleshoot the machines and devices in the specific process plant where they work, which allows them to investigate and fix issues that may arise. Operators do not need to be engineers to verify the abnormal behaviour of processes and operations. They can physically go to the field and check the machines and devices in person per their training procedures or use several tools to troubleshoot systems. Physical malicious acts or equipment failures can be identified by visual inspection.

The AO will be trained to understand how the presented process plant model usually operates and learn the operator troubleshooting procedures. Once a SCADA alarm is triggered, the AO will understand that abnormal behaviour is occurring. The AO will collect data from the CPS to determine what causes the abnormal behaviour, which could be classified as CPA or ordinary faults such as system or device failure.

If the cause of the alarm is deemed a malicious attack, the AO will collect information from all layers of the MPP. When CPA is identified, the AO communicates to the Operator via the SCADA HMI.

2.8 CONCLUSIONS

The effects of cybersecurity concerns and the fear of being hacked on people's mental health have been widely investigated through research. Depression, anxiety, and panic attacks were found to be some of the illnesses and symptoms associated with the pressure of dealing with personal and professional life environments and responsibilities. Additionally, the isolation at the mining site and continuous worries about safety and financial losses may exacerbate the mental health issues experienced by the Operator.

Proposed solutions for autonomous or aware CPS have varied in their approach to the Operator's participation. Some have suggested eliminating the Operator's involvement, while others have sought to minimise it. However, most of these solutions focus on transforming the Operator into a cybersecurity expert, which would be costly and shift their primary responsibility from maintaining the process plant's availability and functionality to monitoring cybersecurity. Conversely, the AO aims to expand the Operator's capabilities without requiring additional knowledge or attention towards cybersecurity.

The research review revealed gaps in the operator's role and mental health state. This study suggests an AO to enhance the operator's capabilities rather than create a cybersecurity expert. The AO will be trained using abnormal operation behaviour, which refers to the logical sequence of expected processes and events in the plant's operations. Research focusing on network traffic monitoring may not effectively capture the Operator's domain knowledge. Therefore, an AO that combines cybersecurity knowledge with the Operator's skillset could be a more suitable solution.

It is important to note that alarms in the SCADA system may result from attacks or fake data injection, leading to false information. Future work should include evaluating the impact of the AO on the Operator's mental health and alleviating the burden of cybersecurity intricacies. Additionally, improvements to the AO's machine learning engine should be made to increase the accuracy of correct CPA identification over false positives. Once the AO is familiar with the MPP's operations and processes, standard metrics will be established for a comprehensive survey and comparison of results.

REFERENCES

Adepu, S., Brasser, F., Garcia, L., Rodler, M., Davi, L., Sadeghi, A.-R., & Zonouz, S. (2020). Control behavior integrity for distributed cyber-physical systems. In 2020 ACM/IEEE 11th International Conference on Cyber-Physical Systems (ICCPS). https://doi.org/10.1109/iccps48487.2020.00011

Australia, G. o. W. (2019, January 3). *PMA20116 certificate II in process plant operations*. Government of Western Australia South Metropolitan Tafe WA.

Retrieved January 3, 2019, from www.southmetrotafe.wa.edu.au/courses/certificate-ii-process-plant-operations

Bada, M., & Nurse, J. R. C. (2020). *The social and psychological impact of cyberattacks* (pp. 73–92). Elsevier. https://doi.org/10.1016/b978-0-12-816203-3.00004-6

Chen, Q., Trivedi, M., Abdelwahed, S., Morris, T., & Sheldon, F. (2016). Model-based autonomic security management for cyber-physical infrastructures. *International Journal of Critical Infrastructures*, *12*(4), 273–294. https://doi.org/10.1504/ijcis.2016.081297

Coates, G. M., Hopkinson, K. M., Graham, S. R., & Kurkowski, S. H. (2010). A trust system architecture for SCADA network security. *IEEE Transactions on Power Delivery*, *25*(1), 158–169. https://doi.org/10.1109/tpwrd.2009.2034830

Do, V. L., Fillatre, L., Nikiforov, I., & Willett, P. (2017). Security of SCADA systems against cyber-physical attacks. *IEEE Aerospace and Electronic Systems Magazine*, *32*(5), 28–45. https://doi.org/10.1109/maes.2017.160047

Gross, M. L., Canetti, D., & Vashdi, D. R. (2016). Cyberterrorism: Its effects on psychological well-being, public confidence and political attitudes. *Journal of Cyber Security*. https://doi.org/10.1093/cybsec/tyw018

Huang, S., Zhou, C.-J., Yang, S.-H., & Qin, Y.-Q. (2015). Cyber-physical system security for networked industrial processes. *International Journal of Automation and Computing*, *12*(6), 567–578. https://doi.org/10.1007/s11633-015-0923-9

Hurst, W., Merabti, M., & Fergus, P. (2013). Behavioural observation for critical infrastructure security support. In *2013 European Modelling Symposium*. https://doi.org/10.1109/ems.2013.6

Inkster, B., Knibbs, C., & Bada, M. (2023). Cyber security: A critical priority for digital mental health [perspective]. *Frontiers in Digital Health*, *5*. https://doi.org/10.3389/fdgth.2023.1242264

Jones, R. A., & Horowitz, B. (2012). A system-aware cyber security architecture. *Systems Engineering*, *15*(2), 225–240. https://doi.org/10.1002/sys.21206

Kriebel, F., Rehman, S., Hanif, M. A., Khalid, F., & Shafique, M. (2018). Robustness for smart cyber physical systems and internet-of-things: From adaptive robustness methods to reliability and security for machine learning. In *2018 IEEE Computer Society Annual Symposium on VLSI (ISVLSI)*. https://doi.org/10.1109/isvlsi.2018.00111

Layal, A. S., Shah, M., Shah, D., & Joshi, K. (2016). Automatic wireless SCADA system. *International Journal of Engineering Trends and Technology*, *34*(2), 81–84. https://doi.org/10.14445/22315381/ijett-v34p217

Li, W., Xie, L., Deng, Z., & Wang, Z. (2016). False sequential logic attack on SCADA system and its physical impact analysis. *Computers & Security*, *58*, 149–159. https://doi.org/10.1016/j.cose.2016.01.001

Metso Corporation. (2018). *Basics in mineral processing* (Metso, Ed., 11th ed., Vol. 1). Author. www.metso.com/contentassets/0efc5d1a7c5a4357baecc5e990dc1fe7/basics-in-mineral-processing-handbook-18-lr.pdf

Nazir, S., Patel, S., & Patel, D. (2017). Assessing and augmenting SCADA cyber security: A survey of techniques. *Computers & Security*, *70*, 436–454. https://doi.org/10.1016/j.cose.2017.06.010

Nicholson, A., Webber, S., Dyer, S., Patel, T., & Janicke, H. (2012). SCADA security in the light of cyber-warfare. *Computers & Security*, *31*(4), 418–436. https://doi.org/10.1016/j.cose.2012.02.009

Noraishah Ismail, S., Ramli, A., & Abdul Aziz, H. (2021). Research trends in mining accidents study: A systematic literature review. *Safety Science*, *143*, 105438. https://doi.org/10.1016/j.ssci.2021.105438

Parreno, J. (2017). What are the different types of mining and how do they differ? *Machines4U*. Retrieved December 28, 2018, from www.machines4u.com.au/mag/different-types-mining/

Pizarro, M. J., & Fuenzalida, A. F. (2021). Mental health in mine workers: A literature review. *Industrial Health*, *59*(6), 343–370. https://doi.org/10.2486/indhealth.2020-0178

Sayan, C., Hariri, S., & Ball, G. (2017). Cyber security assistant: Design overview. In *2017 IEEE 2nd International Workshops on Foundations and Applications of Self* Systems (FAS*W)*. https://doi.org/10.1109/fas-w.2017.165

Services, C. (2018). Control room operator – course overview. *Coal Services*. Retrieved January 3, 2019, from www.coalservices.com.au/wp-content/uploads/2016/11/MR-T_control-room-op_V02_20163006.pdf

Steel, A. (2022, October 27). The psychological impact of cyber attacks. https://blog.lastpass.com/2022/08/the-psychological-impact-of-cyber-attacks/

Training, A. G. (2018). RIIUND308D conduct control room operations. *Australian Government Training*. Retrieved January 3, 2019, from https://training.gov.au/TrainingComponentFiles/RII/RIIUND308D_R2.pdf

Yampolskiy, M., Horvath, P., Koutsoukos, X. D., Xue, Y., & Sztipanovits, J. (2012). Systematic analysis of cyber-attacks on CPS-evaluating applicability of DFD-based approach. In *2012 5th International Symposium on Resilient Control Systems*. https://doi.org/10.1109/isrcs.2012.6309293

Yihdego, Y., & Drury, L. (2016). Mine dewatering and impact assessment in an arid area: Case of gulf region. *Environmental Monitoring and Assessment*, *188*(11), 1–13. http://dx.doi.org/10.1007/s10661-016-5542-6

Chapter 3

Building cognitive resilience for enhanced cyber governance

Marthie Grobler and Tooba Aamir

3.1 INTRODUCTION

In an online world, cybersecurity is often pictured as a computer or a keyboard with a locked padlock and potentially some asterisks to give the impression of data privacy. In phishing-specific imagery, you often see a laptop with a physical fishing rod reaching either in or out of the laptop to hook a credit card or a personal identification card. Cyberthreats are often pictured as a dark hoodie-clad figure hunching behind a laptop or strings of bits and bytes in white, yellow, or green text on a dark background, with the word 'password' in red text. These imageries are all very typical of cybersecurity representation. These imageries are also far removed from how people experience and perceive cybersecurity and cyberthreats.

People generally associate cybersecurity with a password or lock combination, an antivirus programme, and some form of multi-factor authentication. In a simplified form, people believe these measures are sufficient to reach that locked padlock state of cybersecurity. On an organisational level, an added firewall and intrusion detection system may be considered sufficient cybersecurity protection. There is also some kind of disconnect, with people often disassociating cybersecurity with other smart devices such as mobile phones, tablets, smart-home speakers and smart TVs, and believing that cybersecurity is only something that impacts their laptop or desktop. That is one of the primary reasons why people can very easily fall for social engineering attacks because this attack type relies on human interaction and does not fit within the typical hoodie-clad figure imagery.

It is very difficult for humans to comprehend these intangible concepts. In many cases, humans take on the attitude that if they cannot see it, it is not real, so they ignore it. That makes it very easy to exploit human nature and human psychology. The fact that cyberthreats and attacks are ever-changing makes it even more difficult for humans to keep up and easier for attackers to catch somebody off-guard. That is why there is a need to focus on cyber resilience. Humans must be cognitively resilient to govern their cybersecurity successfully – the human brain is built on knowing and understanding, learning and thinking, and *adapting*. It is within human nature to be cognitively resilient, and by channelling this correctly, humans should be able to govern

DOI: 10.1201/9781032664859-3

their cybersecurity efficiently. Therefore, a shift is needed from the adage that humans are the weakest link. Humans become increasingly vulnerable as they merge with technology and cyberspace, exposing themselves to various intangible threats and risks (Grobler et al., 2021). As such, a deeper understanding of human behaviour and behavioural economics will provide an alternative lens on cybersecurity governance. Where cybersecurity solutions often focus solely on technical or operational mechanisms or human-centred solutions, converging technical and operational, strategic and human-centred solutions that overlay user competencies on various skill sets and abilities is much needed to build cognitive resilience for enhanced cybersecurity governance.

Humans need to *know* what they are doing to manage their cybersecurity. Humans need to *understand* what can go wrong, be willing to *learn* from their (and others') experiences, and *think* under pressure to navigate the digital world successfully. With the ability to effectively adjust to challenging experiences, humans can easily *adapt* to the fast-moving digitisation of the world and thrive in a world that requires them to have their cognitive wits about them. This chapter explores how cognitive resilience can be built to ensure enhanced cybersecurity governance. We will examine how cybersecurity attacks and events aim to exploit human psychology in the context of governance and how humans' innate resilience and cognitive abilities can help them navigate these events. We will look at triggers that adversaries can manipulate to breach security and at humans' understanding of complex connections that can help them to think under pressure. We will discuss how these tactics compromise decision-making processes, data sharing, and citizen engagement and could contribute to the potential erosion of public trust, compromise sensitive data, and disruptions in governance operations. We will investigate the role and effectiveness of awareness campaigns that harness psychological insights to empower citizens or executives with the knowledge and skills to recognise, resist, and report social engineering attempts. Simply put, we will explore how humans think and act under pressure and how this impacts their cybersecurity standing, either positively or negatively.

3.2 HUMAN PSYCHOLOGY AND COGNITIVE ABILITIES

Psychology is the scientific study of the human mind and behaviour. It explores how we perceive, remember, and communicate and investigates complex psychological disorders and cultural influences on cognition. The study of human psychology and cognitive abilities is a fundamental branch of psychology that deals with the intricate workings of the human mind and behaviour. It focuses on understanding the underlying mental processes and mechanisms that govern human cognition. Cognitive psychologists employ experimental methods to unravel these facets of human cognition, and integrating cognitive neuroscience facilitates a deeper understanding of the neural basis of cognitive functions.

The significance of cognitive psychology extends to advancing our understanding of human behaviour. It provides invaluable insights into the cognitive processes that steer our actions, decisions, and interactions with the world. Beyond its foundational importance, cognitive psychology has far-reaching implications, contributing to various branches of psychology, including clinical psychology, developmental psychology, and social psychology (Núñez et al., 2019). This multidisciplinary relevance underscores the imperative nature of comprehending the intricacies of human cognition as it finds applications in diverse fields, ranging from educational methods and the treatment of cognitive disorders to enhancing user experience in technology design.

Originating in the mid-20th century, cognitive psychology has profoundly impacted our understanding of how individuals perceive, think, learn, and ultimately make decisions (Fleming, 2019). Cognitive psychology embodies several vital principles and concepts that serve as the cornerstones of this discipline. Specifically, it considers how humans *process information*. Cognitive psychologists often adopt the information processing model, which draws a parallel between the human mind and a computer. This model speculates that the mind receives information, processes it, and generates responses based on this processing. It considers the ability to *retain and divide attention*. It considers how individuals allocate their mental resources to various stimuli in their environment and the role of attention in the *perception* of the world, short-term and long-term *memory* and subsequent forgetting. Cognitive psychology plays a sizeable role in *language and communication*, including language structure and comprehension. Finally, it considers *problem-solving*, decision-making, and how people navigate challenges (O'Reilly & Munakata, 2000).

The principles and concepts within this field form the basis for our understanding of human cognition, with far-reaching applications in various domains of science and practical endeavours. In essence, cognitive psychology represents an exploration of the cognitive processes that define our interactions with the world, shaping our thoughts, actions, and decisions.

3.3 DIGITAL WORLD AND CYBERSECURITY GOVERNANCE

Digitisation is complicating our understanding of human behaviour. People cannot keep up with technological changes, and their main focus is often not on cybersecurity *per se* but instead on their primary goal. For example, somebody using a search engine focuses on finding more information on what they are looking for, not the potential cyber dangers. Similarly, somebody using a mobile phone to navigate the train timetable focuses on getting from point A to point B and not looking for potential malicious website redirection.

The rapid digital transformation of the current economy and society is causing significant cybersecurity gaps that increase the risk associated with cyber incidents across different economic areas. It is, therefore, necessary to upskill and upscale cybersecurity governance in this digital world.

Cybersecurity governance is a comprehensive strategy that integrates critical functions and supporting enablers to prevent and manage cyberthreats and cyberattacks. Building on the NIST (2023) cybersecurity framework, functions may include identification, protection, detection, response, and recovery-related actions. Examples of these functions include identifying digital assets, their risk profiles and current cybersecurity posture, and the *protection* of these assets by implementing appropriate risk mitigation mechanisms and system hardening, either through technical or procedural changes (NIST, 2023). Functions may include *detection*-specific activities, such as log analysis and network scanning, to identify any anomalous device or network behaviour; *response* functions may include immediate remediation activities to contain any threats and limit the immediate consequences, whilst *recovery*-related activities focus on more long-term consequences and impact management. A cybersecurity governance framework is needed to ensure adequate consideration is given to covering the entire spectrum of the organisation's cyber needs through a single integrated framework that separates cyber governance from cyber management while meeting stakeholder needs and enabling data sharing (ISACA, 2012).

These functions are supported by enablers that provide the structure and support for the organisation to implement the governance functions. The enablers include policies, frameworks, legislation, processes, organisational structures, culture, ethics, behaviour, information, services, infrastructure, applications, people, skills, and competencies (ISACA, 2012). *Policies, frameworks*, and *legislation* may include organisational information security policies, national-level legislation such as the Australian Privacy Act 1988, and international frameworks and best practices such as the ISO/IEEE 27000 series. *Processes* may be very organisation-specific, such as detailing the steps for user authentication or how to report a stolen or misplaced laptop. *Organisational structures* may include the constitution of a board committee or an information security steering committee, appointing a Chief Information Security Officer, or using an external security consulting service. *Culture*, *ethics*, and *behaviour* relate to top-down cybersecurity enforcement, bottom-up staff involvement, the visibility of an active cybersecurity champion, and staff's attitude to cybersecurity-related reporting. *Information* is the basis of communication and critical to process flows and operationalisation. *Services*, *infrastructure*, and *applications* are the hardware, firmware, software, and middleware needed to process and commoditise the information. Finally, *people* refer to the users and beneficiaries of the information and services with the *skills and competencies* needed to run the business and cybersecurity functions.

Cybersecurity governance is needed to create structure and provide tangible elements in an otherwise chaotic, intangible cyber domain. A cybersecurity governance framework is a mechanism used to help humans manage cybersecurity in a way that aims to bridge some of the gaps created by digitisation and upscale cybersecurity management in an otherwise non-cyber focused organisational environment. A cybersecurity governance framework can be customised to characterise and evaluate the maturity level of organisations across critical domains. Although cybersecurity functions are traditionally only used in large organisational settings, the basic principles are the same regardless of organisational context, size, or sector. They can be adapted to any maturity level. Such a framework can be used to evaluate the organisation's current and aspirational target cybersecurity profiles according to the framework. It can be applied to increase the cybersecurity culture, set the priorities of investment in information security, and measure the implementation progress. It can also identify and evaluate latent cybersecurity risks, including understanding the probability of occurrence and possible impact on the organisation's operational continuity.

A cybersecurity governance framework is a risk-based approach to managing cybersecurity risks in an increasingly digitised world. The framework provides a common language for understanding, managing, and expressing cybersecurity risk to internal and external stakeholders. It can be used to help identify and prioritise actions for reducing cybersecurity risk, and it is a tool for aligning policy, business, and technological approaches to managing that risk. Such a framework intends to understand cybersecurity through a methodical business lens (particularly through the governance functions that categorise security aspects into logical blocks of actions). Similarly, the governance framework facilitates a better understanding of the business functions and operations. This is done through the system mapping and interconnectivity processes uncovered during the specific cybersecurity functions' activities and the framework enablers' structures and processes. Implementing a cybersecurity governance framework can support the development of cognitive resilience. Appreciating human behaviour affects governance and can lead to enhanced cybersecurity.

3.4 SOCIAL ENGINEERING AND HUMAN PSYCHOLOGY

Social engineering, a sophisticated criminal technique, manipulates individuals to reveal confidential information or perform actions contrary to their usual behaviour (Hadnagy, 2010). This section delves into the psychological underpinnings of social engineering, crucial for recognising and defending against its tactics, thereby enhancing resilience to manipulation by malicious actors. The discussion encompasses a range of deceptive techniques exploiting human psychology and trust, emphasising the exploitation of vulnerabilities rather than relying solely on technical exploits.

The effectiveness of social engineering hinges on its exploitation of human psychology, with malicious actors understanding and manipulating cognitive and emotional responses to achieve their objectives (Uebelacker & Quiel, 2014). Fundamental principles influencing human behaviour and trust, such as influence, persuasion, authority, and cognitive biases, play a pivotal role. Psychological principles like reciprocity, commitment, social proof, authority, and scarcity are utilised to manipulate targets through *influence and persuasion*, creating situations where compliance with requests or *trust in authority* is expected. This section also explores social engineers' specific exploitation of trust and authority, including impersonating trusted figures or entities and manipulating *cognitive biases* to align information with targets' existing beliefs.

The evolving landscape of social engineering, adapting to changing technologies and circumstances, makes it a pervasive cybersecurity threat (Wang et al., 2021). Unlike traditional computer attacks, social engineering focuses on exploiting human vulnerabilities, allowing attackers to bypass security measures without directly engaging with technical defences. The expansion of the cyber environment, including avenues like email, mobile phones, and the Internet of Things (IoT), further intensifies the social engineering threat. Open office environments, BYOD policies, and remote work have expanded the attack surface for social engineers (Wang et al., 2020).

The availability of open-source intelligence simplifies information gathering and facilitates crafting more credible and targeted social engineering attacks (Abraham & Chengalur-Smith, 2010). Prevalent techniques, including pretexting, impersonation of authority figures, and fabricating emergencies, capitalise on human vulnerabilities, creating avenues for attacks. Integrating technologies like machine learning and artificial intelligence are expected to enhance the efficiency and aggressiveness of social engineering attacks, making targeted, large-scale, and automated attacks a genuine possibility. Thus, social engineering has evolved into a persistent and universal security threat.

In an era marked by technological progress, the human element in securing digital landscapes gains significance. Cognitive psychology, dedicated to exploring mental processes, emerges as a pivotal force in reshaping cybersecurity strategies (Pfleeger & Caputo, 2013). This transformative role extends beyond technological alignment, forming a foundational framework for anticipating and countering cyberthreats. Integrating cognitive psychology into cybersecurity emphasises human factors, delving into human vulnerabilities, trust dynamics, and cognitive biases.

This section further examines how cybersecurity attacks exploit human vulnerabilities through psychological approaches, impacting trust dynamics and cognitive biases. For instance, cognitive psychology provides insights into how individuals may succumb to phishing attacks due to innate tendencies, like the inclination to trust familiar sources. Recognising these vulnerabilities is crucial in designing cybersecurity measures that shield against external threats while accounting for inherent weaknesses in human cognition. A human-centric approach places skilled individuals at the core of security concepts, with technology supporting human capabilities throughout the entire technology life cycle.

The decisions made by individuals, whether end users or cybersecurity professionals, profoundly impact the security posture of digital environments. Cognitive biases, information processing models, and external stimuli's influence on decision-making become critical analysis points in crafting effective cybersecurity strategies. The human-centric approach advocated by cognitive psychology addresses the multifaceted challenges presented by human vulnerabilities, trust dynamics, and cognitive biases. Embracing a multidisciplinary perspective that integrates insights from cognitive psychology enhances the technical robustness of security measures and strengthens cybersecurity's adaptive and resilient nature. This approach acknowledges the constant interplay between human factors and technological advancements within the evolving governance context, providing a cornerstone in crafting strategies that navigate the intricate interplay of human cognition in the ever-evolving digital landscape. To illustrate the relationship between human vulnerabilities, trust dynamics and cognitive biases, consider the following examples.

3.4.1 Example 1: malicious actors use whaling attack to defraud bank of US$75.8 Million

In 2016, malicious actors targeted a high-ranking executive of the Belgian Bank Crelan with a spear-phishing email. The attack focused on exploiting weaknesses in the institution's internal communication channels and employee training programmes. The malicious actors created convincing scenarios that prompted employees to disclose sensitive information, including login credentials and financial transaction details. The malicious actors accessed sensitive financial information and transaction records by exploiting a vulnerability in the institution's email system. This breach exposed weaknesses in the organisation's cybersecurity infrastructure, revealing gaps in employee training, communication protocols, and system vulnerabilities. None of the bank's clients were affected, and the bank's reserve covered the loss (Cimpanu, 2016).

Human Vulnerabilities: The malicious actors targeted the organisation's internal communication platforms and financial databases, using a combination of pretexting and impersonation to deceive employees. They relied on information processing by the targeted individual and influenced their perception of their immediate surroundings and decision-making by leveraging urgency and perceived authority to create a stressed environment that required immediate action.

Trust Dynamics: By impersonating high-ranking executives and employing tactics such as authority and urgency, the malicious actors gained access to critical financial systems. The email used legitimate graphics and a lookalike domain name to fool the targeted individual and have them transfer the money without double-checking with somebody from inside the organisation first.

Cognitive Biases: The malicious actors leveraged cognitive biases, such as trust in authority figures, to convincingly pose as legitimate personnel, making it challenging for employees to discern the malicious intent. The malicious actors strategically manipulated these biases to navigate the organisation's security measures, emphasising the critical role of understanding human behaviour in cybersecurity.

3.4.2 Example 2: malicious actors use AI voice impersonation to defraud organisation of US$243,000

In 2019, malicious actors used AI deepfake technology to generate a voice impersonation of a chief executive to social engineer the CEO of a UK-based energy organisation. The malicious actor, pretending to be the CEO's boss, demanded an immediate transfer of US$243,000 to a supplier. Since this was outside the standard operational guidelines, the impersonator falsely assured the CEO that the amount would be reimbursed. After the money was transferred to the specified Hungarian-based supplier, the money was transferred again to a Mexico-based account and then to further locations, making identifying the malicious actors and tracking the funds more difficult. The malicious actors made second and third attempts to engage the original victim, but these attempts were unsuccessful (Trend Micro, 2019).

Human Vulnerabilities: The CEO was easily influenced and persuaded by the voice impersonation of their boss. It was reported that the CEO 'recognised the subtle German accent in his boss's voice – and that it carried the man's "melody"' (Damiani, 2019). People are inclined to readily accept direct orders from their line managers as risk acceptance is perceived on a higher authority level. The malicious actors employed several social engineering techniques, including pretexting, impersonating an authority figure, and fabricating an emergency or critical situation.

Trust Dynamics: The CEO trusted the chief executive's authority as a senior figure within the organisation. Initially, there was no distrust; therefore, the CEO felt no need to verify the caller's identity. This trust was broken after the transferred funds were not reimbursed, leading to unsuccessful second and third attempts to social engineer the same individual.

Cognitive Biases: The CEO believed that the voice impersonation represented their boss, influencing their response to the out-of-procedure funds transfer request, intending to support their boss and appear to be a good employee. The malicious actors leveraged the perceived risk compensation bias to build trust with the targeted individual.

These examples demonstrate different social engineering techniques using different technology platforms but intending to leverage a perceived authoritative trust to persuade unsuspecting individuals to forego the proper procedure. Although not specified in the examples, it can be assumed that both organisations had some procedure stipulating the authorities needed to transfer funds. However, since the human element was targeted, none of these procedures were presumably followed since various cognitive factors unduly influenced the humans' decision-making. For example, the employee may have wanted to impress the boss by actioning any direct tasks immediately and effectively. The employee may have assumed that the boss would not suggest anything illegal or violate the organisational policy and assumed that any direct tasks would be acceptable. If the employee had any reservations about the process, they may have assumed risk transfer in that the order came from a more senior individual, and therefore the more senior individual would be responsible for any adverse outcomes. Social engineers can easily identify possible cognitive biases through reconnaissance of a targeted organisation. In addition, social engineers often develop multiple pretext scenarios that will enable them to manipulate individuals regardless of their engagement, decision-making, or trust levels. In these cases, the impact of these social engineering attacks was profound, resulting in significant financial losses and reputational damage, both for the organisation and the involved individual. The impact of these attacks was not limited to immediate financial losses but also extended to the erosion of trust among customers and stakeholders. Regarding the targeted individual, various factors would contribute to how their experience would shape their future cyber resilience. In the ideal environment, this individual would become a cybersecurity champion who shares their story and advocates for better cybersecurity governance that considers human behaviour. However, in some cases, the individual may be ostracised and shunned from an organisational perspective, affecting their cybersecurity standing to such an extent that they become entirely disengaged.

3.4.3 Example 3: malicious actors exploit human nature to defraud australians of more than AU$7.2 Million

In 2022, Australians have lost more than AU$7.2 million in family impersonation text scams. The malicious actors generally send a message, usually via WhatsApp or SMS, posing as a friend or family member who has lost their phone or got a new number. There are different variations, but the malicious actors generally fabricate a story related to them being locked out of the bank account and asking the victim to send money to them. According to the Australian Competition and Consumer Commission, more than 11,000 incidents were reported in Australia in 2022. Although any mobile number can receive these fraudulent texts, the scam targets women and older individuals (Bahr, 2023).

Human Vulnerabilities: This attack is grounded purely in human nature and banking on the fact that a mum would do whatever she could to help her child. It also leverages the failures of human–technology interaction, as many people would have had to borrow somebody else's mobile phone in a pinch to send a message or make a call in an emergency.

Trust Dynamics: The victim trusts that their family or friends will reach out to them when in need. Given the demographics of the targeted victims, these individuals are not necessarily considered tech-savvy and will automatically assume that the message is legitimate and would not consider verifying the texter's identity.

Cognitive Biases: Over two-thirds of family impersonation scams have been reported by women over 55 years of age, so scammers are combining the fight or flight adrenaline rush with familial behavioural patterns and exploiting the fact that people will step up to protect their families.

This example highlights that human vulnerabilities can be exploited in any circumstance and are not restricted solely to work environments. Cognitive biases are present in all levels of mental reasoning, regardless of whether it is a work or personal environment. Trust dynamics are equally prevalent in both environments, specifically personal ones, since human relations are often more private and individualised. Although the financial impact is often insignificant in the bigger scheme of things, falling victim to such a scam could be devastating for an individual due to the loss of money and the perceived breach of trust relationship with the impersonated individual. In addition, victims of these scams often develop new negative cognitive biases against the benefits of technology and often never regain complete trust in the digitised world again.

3.5 PUBLIC TRUST AND CITIZEN ENGAGEMENT

Exploring the implications for public trust and citizen engagement is essential in social engineering. The tactics utilised in social engineering attacks can potentially undermine decision-making processes, compromise the sharing of critical data, and disrupt the level of citizen engagement. This, in turn, directly threatens the functioning of governance operations and the credibility of public institutions. Successful social engineering attacks can erode public trust, leading to increased scepticism and reduced engagement. Additionally, compromising sensitive data jeopardises individuals and organisations and threatens the confidentiality and security that citizens expect from the relevant organisations. These issues underscore the need for robust security measures and public awareness to protect public trust and citizen engagement in an increasingly prevalent era of digital threats. This section will discuss how these tactics compromise decision-making processes, data sharing, and citizen engagement.

Cybersecurity incidents, especially those that occur on a large scale, significantly impact the reputational perception of the involved organisation. Not only will the users directly affected by the incident be negatively associating the organisation with their financial losses and changed social standing, but through cognitive biases, these negative associations are often transferred to other individuals whom the specific cybersecurity incidents may not have directly impacted. As described in the Situational Crisis Communication Theory, the attributions of crisis responsibility significantly affect how people perceive an organisation's reputation in crisis and their affective and behavioural responses to that organisation following the crisis (Kuipers & Schonheit, 2021). Through social connectedness, people often empathise with the negative experiences shared by others and are accordingly influenced by these negative impacts. Furthermore, the power of social media sentiment is a powerful tool that can influence people towards a specific state of trust or mistrust in an organisation. It, therefore, remains an evolving challenge to actively engage citizens and build their trust in cybersecurity, especially if previous cyber incidents have broken their trust. To illustrate the relationship between public trust and citizen engagement, consider the following examples.

3.5.1 Example 4: authentication gaps: losing user trust to malicious exploits

In 2022, malicious actors breached Optus's security, affecting up to 10 million current and former customers and over a third of the Australian population. The breach was effected through an application programming interface (API) available online with no authorisation or authentication required – anybody with the URL could access the data (Taylor, 2022). This breach resulted in a lawsuit that claims Optus breached consumer and telecommunications law and failed in its duty of care to protect users from harm.

Public Trust: Optus claimed in their media statements that a sophisticated attack caused the data breach. However, it has been shown that the lack of authentication controls on the API was a very trivial cybersecurity omission. Optus' lack of care in basic cybersecurity measures significantly damaged public trust. They acknowledged that the public's trust in the organisation had been broken and confirmed that an independent external review was crucial in rebuilding customer trust (*ABC*, 2022a).

Citizen Engagement: The incident and Optus' handling of the event caused a lot of mistrust and disengagement amongst customers, as evidenced by negative sentiments on social media accounts and other public forums. One month after the cyber incident, 10% of affected users left the network for competitor networks, with an additional 56% considering a change (Sharples, 2022). Optus has been criticised for handling the incident, and users have been very outspoken about their future engagement with the organisation.

3.5.2 Example 5: credential stuffing, blaming users for data theft

In 2023, malicious actors reused old usernames and passwords from other websites (known as credential stuffing) to break into DNA testing organisation 23andMe's database, stealing the personal information of nearly 7 million customers. The malicious actors used this information to find the impacted customers' genetic relatives and family tree information through an additional opt-in feature (Lyngaas, 2023). Following the breach's discovery, 23andMe instructed affected users to change their passwords and later rolled out two-factor authentication for all its customers (MacDonald, 2023). 23andMe is blaming wholly on the customers who negligently used recycled username and password combinations as login credentials for their accounts on the organisation's website. Their stock dropped more than 8% after this announcement (Revell, 2024).

Public Trust: Although some of the user accounts were breached due to users' poor password hygiene, many additional user accounts were accessed due to the additional opt-in feature selected by some users – these accounts were not breached due to poor password hygiene. Given 23andMe's response to the current lawsuit by plaintiffs affected by this breach, the public trust in the organisation has been broken. The organisation also only instituted two-factor authentication after the incident; considering the sensitive nature of the data they access, this should have been an essential security measure. The possibility exists that personal trust relationships are impacted due to familial connections identified inadvertently by the breached data.

Citizen Engagement: Given the personal nature of the data loss (including genetic markers and familial data), it is expected that citizens affected by this breach would become very disengaged in pursuing online interactions. Any digital aspect related to biometric factors would no longer be considered secure for these individuals. There is a possibility that affected users would opt for better password hygiene in future, not reusing passwords, possibly scrutinising the security of other websites where they have reused passwords and implementing two-factor authentication on more accounts.

These examples highlight that cybersecurity incidents often negatively impact public and personal trust relationships. In terms of citizen engagement, the companies involved in the breach may experience a sharp reduction in citizen engagement, whilst the affected users might inversely become more engaged in their cybersecurity considerations for other platforms and services. In addition, the two examples show diverse public framing of the incidents and the impact thereof on the organisation's reputation. Framing relates crises to different degrees of attributed responsibility. It corresponds to minimal crisis responsibility (victim crisis is outside of the organisation's control, such as natural disasters or employee misbehaviour), low-moderate crisis responsibility (accidental crisis through an unfortunate series of events), and high crisis responsibility (preventable crisis where the organisation either intentionally caused the crisis or could have avoided the occurrence but failed to do so). The more an organisation is perceived to be accountable for a crisis, the more its reputation will suffer, as it will no longer be trusted to adequately prevent or respond to crises (Kuipers & Schonheit, 2021). In Example 4, the organisation claimed a sophisticated attack, which was a fundamental security omission. In Example 5, the organisation turned on their customers and shifted all the blame. In both instances, the organisations could have better utilised an understanding of cognitive psychology to guide their use of language and communication in a way that would have influenced their customers more positively towards them and actively re-engaged them whilst building trust. In addition, a cybersecurity governance model might have supported them in better handling the consequence management of the events and the subsequent recovery processes.

3.6 CYBER AND COGNITIVE RESILIENCE BUILDING

To adequately prepare for and address cybersecurity incidents, the focus should shift towards the critical cyber and cognitive resilience domains within cybersecurity. This section explores how cognitive resilience can build a cyber immunity that can be adopted in modern cyber governance frameworks. We investigate cognitive resilience that can be strategically incorporated into modern cyber governance frameworks, allowing for a proactive approach to mitigate the ever-evolving cyberthreats that confront organisations and individuals and establish robust cyber immunity. By assessing these dimensions, we aim to underscore the pivotal role that cognitive resilience plays in enhancing cybersecurity strategies and fortifying defences against the multifaceted challenges posed by cyber incidents in the digital landscape.

Within the human-centric cybersecurity domain, the focus is on all things of cybersecurity where humans are involved, particularly on human involvement in the system and processes. That is, understanding how humans represent value but risk to an organisation, how humans and computers interact, and what risks are introduced due to these interactions. This human-centric cybersecurity focus helps build cognitive resilience by looking at cybersecurity with a behavioural lens and considering cybersecurity's user, usage, and usability aspects (Grobler et al., 2021) as a means towards true cyber and cognitive resilience. It is part of human nature to be cognitively resilient, and by channelling this correctly, humans should be able to govern their cybersecurity efficiently. Therefore, a shift is needed from the adage that humans are the weakest link. The convergence of technical and operational, strategic and human-centred solutions that overlay user competencies on various skill sets and abilities is much needed to build cognitive resilience for truly enhanced cybersecurity governance. Therefore, we must leverage the human brain's ability to *understand*, *learn*, *think*, and *adapt*. These metacognitive abilities support the cybersecurity governance framework's enabler for *people*, *skills*, and *competencies*.

3.6.1 Humans need to know what they are doing to manage their cybersecurity

As mentioned earlier, people cannot keep up with the changes in the digital environment and often, their main focus is not on cybersecurity *per se* but instead on their primary goal. Although, understandably, not all technology users are experts in deep technology mechanics, there should be an assumption that all technology users have a basic level of cyber literacy. For example, there would be an expectation that all email users are aware of basic phishing techniques and would be able to identify blatant phishing attempts that include poor spelling and grammar, an obviously fake logo, and the lack of a personalised greeting in an email. With advances in technology, social engineering attempts have also significantly improved, and therefore, there are now more targeted attacks, as illustrated in Examples 1, 2, and 3. In these examples, the user should have their wits about them to identify minute but nuanced differences from what is expected.

Users need to understand what is expected of them to protect themselves and manage their cybersecurity and not fall for elementary cybersecurity attacks. For example, certain demographical groups are very trusting and generally trust whatever is on the internet. If they see a pop-up that says *Your PC is infected, click here to run malware cleaner*. These individuals will click on it with a trusting nature, but instead of ridding them of the malware, they will unintentionally install it. Another example was illustrated in Example 5, where many customers reused the same username and password combination to protect personal genetic data. Cybersecurity advocates have long warned internet users not to follow these practices and to improve their cyber hygiene. A further example, particularly relevant in the organisational setting, is the social engineering variation that uses the organisation's structure to pressure staff by impersonating authority figures. Although these attempts are often factually correct in terms of content, users should be able to pick up nuances; for example, something simple, such as the way the impersonated person usually greets you or the salutation that is different from their regular writing style, can give a phishing attempt away. However, the user must be vigilant and concentrate on the email, or they will miss it. The same goes for a slightly modified email address. If a user is on a mobile phone view of the email app, it would be easy to miss this, but an expanded desktop email will show that the email address looks different than expected.

Particularly for individuals working in a large organisation with a dedicated IT or cybersecurity team, knowing how to manage their cybersecurity is often a significant problem since they are used to somebody else taking care of it. In a user study conducted by CSIRO in 2019, almost 45% of respondents indicated that they rarely change the password for the internet devices used at home, with a further 28% of respondents stating that they only change these passwords when prompted by the associated software. The survey respondents overwhelmingly stated that they do not consider themselves applying the same level of care in terms of cybersecurity in their home environment as opposed to their place of work or study (CSIRO, 2019). Generally, individuals are not familiar with basic cybersecurity measures and do not apply them in their personal environments. Within organisational settings, the dedicated IT or cybersecurity would be more knowledgeable of what measures are in place, but the details of these measures often do not trickle down to the end users or up to the executives that they can comfortably state that they know what they are doing to protect themselves in cyberspace. The problems come in when these individuals access their organisational accounts and networks from a private mobile phone or home network that is not secured correctly.

3.6.2 Humans need to understand what can go wrong

As mentioned earlier, it is very difficult for humans to comprehend intangible cybersecurity concepts, and it is equally challenging to marry deep technology and cybersecurity applications in a non-cyber business environment. It is for this reason that people still fall for phishing emails that leverage cognitive biases (Example 1), cannot easily distinguish between real and deepfake AI-generated content (Example 2), inherently trust what they see (Example

3), claim 'sophisticated attacks' when a cybersecurity omission causes significant disruption (Example 4), and practice poor cyber hygiene (Example 5). To help humans fully understand what can go wrong in the cyber environment, it is necessary to apply proper risk management.

As with any risk, humans can manage cyber risks and cybersecurity incidents in four ways. *Risk avoidance* occurs when users ignore the things that they do not understand. This links to the initial statement that humans often take on the attitude that if they cannot see it, it is not real, and they ignore it. The website https://haveibeenpwned.com illustrates this by showing that the word 'password' has been used as a password in breached accounts more than 9.65 million times, with '123456' used more than 37.6 million times. People do not take password security seriously and enter a basic string to satisfy the password requirements without considering the security aspect. *Risk acceptance and sharing* is another strategy that have mixed success in the cyber environment. Particularly in larger organisations, executives have a cognitive bias that cybersecurity is 'an IT problem' and does not relate to the rest of the business. Without the proper backing of the board of directors, cybersecurity is never taken seriously in these organisations. *Risk transfer* is a strategy where the responsibility for risk mitigation is transferred to a third party. This can be an external cybersecurity consultant, but in many cases, organisations transfer their risk by assuming that their suppliers and third-party vendors have sufficient cybersecurity measures in place. Unfortunately, numerous examples of third-party breaches enabled malicious actors to gain access to a system or network because the third-party cybersecurity measures were not up to scratch. The last strategy is *risk mitigation*, which takes on the form of appropriate cybersecurity governance, including well-developed cybersecurity functions supported by enablers. Ideally, all cybersecurity risks should be handled through proper risk mitigation, where organisations and individuals understand what can go wrong and adequately act to remediate the risk factors before adverse events occur. Proper risk management and understanding what can go wrong play a prominent role in a correctly set cybersecurity governance framework.

3.6.3 Humans need to be willing to learn from their (and others) experiences

Cybersecurity incidents can significantly impact organisations, affecting their economic, social, and organisational dimensions in traumatic and far-reaching ways. The *Cybercrime in Australia* report surveyed 13,887 computer users in early 2023. In the 12 months before the survey, 27% of the respondents had been victims of online abuse and harassment, 22% of malware, 20% of identity crime and misuse, and 8% of fraud and scams. Alarmingly, more than half of the respondents experienced more than one type of cybercrime in the 12 months before the survey (Voce & Morgan, 2023).

There are too many variations of cyber incidents and changes in the digital environment for humans to remain ahead of cyberthreats. Not only is it just the sheer number of attacks constantly increasing, but the variations and inclusion of new technologies and approaches in new domains make the cyberthreat space very lucrative. Therefore, it is necessary for humans to acknowledge and learn from their mistakes and also learn from the mistakes of others. If humans do this, they will be in the best position to understand malicious human behaviour, and behavioural economics will provide an alternative lens on cybersecurity governance. Where cybersecurity solutions often focus solely on either technical or operational mechanisms or human-centred solutions, converging technical, operational, strategic, and human-centred solutions that overlay user competencies is much needed to build cognitive resilience for enhanced cybersecurity governance.

3.6.4 Thinking under pressure: navigating the digital world

Malicious actors hook onto news events; they exploit people's fear and curiosity to get them to click on links and attachments they would have steered clear of if they had thought rationally about it. So, these malicious actors are intent on exploiting human behaviour. With so many things happening in the world around us, it is no wonder that individuals are struggling to connect the real world with malicious online attacks. This is illustrated time and time again when malicious actors adapt their approach to link to a topical issue. Depending on the topic and how interested people are in this, malicious actors can easily exploit human interest and empathy to generate increased traffic to malicious websites and expect increased click-through rates on phishing and scam emails. For example, malicious email traffic increased by almost 600% during COVID-19, cyber-physical attacks rose in the Russian–Ukraine war, Queen Elizabeth II's passing was used in phishing as well as clickbait attacks, and malicious actors used victims' personal information during national disasters to steal money (*ABC*, 2022a, 2022b; Associated Press, 2020; Paganini, 2022; RiskOptics, 2022).

To manage this, humans must think under pressure and see warning signs of malicious activity in new and novel formats to ensure they do not fall victim to apparent scams. In a cross-national study focusing on developing a five-nation phishing resilience benchmark, particular attention was given to cultural and demographical factors that contributed to participants being able to identify malicious indicators under stressful circumstances. Five dimensions were the most significant for impacting people's behaviour in accurately identifying phishing attempts. Specifically, people are more focused on identifying attacks if they perceive that the risk of becoming a cybercrime victim would have a short-term impact (immediacy) if they perceive the risk of becoming a phishing victim as commonly encountered and close to home as opposed to something happening to a friend of a relative (common dread), and if they perceive the severity of becoming a phishing victim as significant (severity) (Gopavaram et al., 2021). Understanding these cognitive biases and human behaviour will enable organisations to more effectively manage awareness campaigns to support their employees in recognising, resisting, and reporting social engineering attempts.

3.6.5 Humans need to adapt to the fast-moving digitisation of the world

Despite ongoing cybersecurity awareness campaigns and training, evolving malicious approaches remain effective globally. Social engineering poses a significant challenge, necessitating users to leverage their adaptability in a world demanding cognitive awareness. A shift from usage-centred to user-centred cybersecurity is crucial, moving from functional systems to those interacting with and supporting users. This fosters a better understanding of human behaviour, dispelling the stigma of users as the weakest link in a cyber system (Grobler et al., 2021).

In the dynamic cybersecurity landscape, strategic use of awareness campaigns emerges as a powerful defence against social engineering attacks. This investigation explores campaigns' pivotal role and effectiveness, utilising psychological insights to empower citizens and executives. The primary objective is to equip individuals with the knowledge to recognise, resist, and report social engineering attempts. Cyber awareness campaigns that integrate psychological insights better uplift cybersecurity ecosystems. Effective management, considering psychological and human factors, enables a successful approach to enhancing resilience against multifaceted cybersecurity incidents. Lessons from the organisational setting should extend to personal environments, empowering users in all digital interactions.

3.7 CONCLUSION

There seems to be a disconnect between the usage of technology, its usability, and the people using it. For the most part, humans do not understand what technology is supposed to do and not what they are capable of doing. Technology is not always that intuitive to use and does not adapt to the needs and tech-savviness of the human. In addition, human nature is complex, and where humans are involved, you need to deal with emotions and agendas, memory, decision-making, and trust. This often gets very messy very quickly. This combination of the user, usage, and usability triad that is not entirely up to scratch introduces weaknesses in the system. To remedy this, organisations must focus on educating their users about cyber hygiene, upskilling their users in correct system usage and educating their management and executives in viewing business objectives with a cyber lens to ensure that their cybersecurity strategy fully covers their fiduciary cybersecurity obligations.

Understanding that humans, in combination with systems and technology, including misuse and misunderstanding of technology in an ever-changing cybersecurity landscape, create a complicated environment where human cognitive abilities are tested and stretched to their limits in adjusting to technological changes and responding to cybersecurity incidents. Therefore, humans need to be aware of the cybersecurity environment in which they find themselves to apply critical thinking and cognitive understanding. Equally, executives of organisations and system designers and implementers need to be aware of human cognition and vulnerabilities, as well as the relationship between these elements with trust dynamics and citizen engagement, to ensure that they can adequately enable users to protect the integrity of digital systems and data.

Understanding that humans alone cannot be solely responsible for the security of systems is essential. Cybersecurity system designers focus only on system aspects to defend against malicious attacks. However, users of cybersecurity systems have diverse perceptions, knowledge, and experience about security risks that drive their behaviours. The failure of any human actor, including system architects, designers, developers, and end users, can cause significant challenges in identifying phishing attacks. Therefore, it is necessary to have a technology system that can double-check and analyse where human errors can occur. This understanding will not only help improve the overall cybersecurity posture of humans but also reduce the impact of and recovery time after cybersecurity incidents.

3.8 ACKNOWLEDGEMENT

This chapter has been supported by the Government of Western Australia acting through the Department of Premier and Cabinet, Office of the Digital Government, Western Australia, and the Cybersecurity Cooperative Research Centre Limited, whose activities are partially funded by the Australian Government's Cooperative Research Centres Programme.

REFERENCES

ABC News. (2022a, October 3). Optus reveals more than 2 million customers had personal ID numbers compromised in cyber attack. www.abc.net.au/news/2022-10-03/optus-data-breach-cyber-attack-deloitte-review-audit/101496190

ABC News. (2022b, February 24). Ukraine isn't just facing the threat of Russian troops on the ground. Cyber attacks are 'intensifying'. www.abc.net.au/news/2022-02-24/ukraine-cyber-attacks-looming-full-scale-russian-invasion/100856586

Abraham, S., & Chengalur-Smith, I. (2010). An overview of social engineering malware: Trends, tactics, and implications. *Technology in Society*, *32*(3), 183–196. https://doi.org/10.1016/j.techsoc.2010.07.001

Associated Press. (2020, May 26). *Top UN official warns malicious emails on rise in pandemic.* www.securityweek.com/top-un-official-warns-malicious-emails-rise-pandemic/

Bahr, J. (2023, January 23). *'Hi mum' scam arrest: What you need to know about text message scams.* www.sbs.com.au/news/article/hi-mum-scam-arrest-what-you-need-to-know-about-text-message-scams/cuh2lcg4b

Cimpanu, C. (2016, January 25). Belgian bank loses €70 million to classic CEO fraud social engineering trick. https://news.softpedia.com/news/belgian-bank-loses-70-million-to-classic-ceo-fraud-social-engineering-trick-499388.shtml

CSIRO. (2019). Population location and movement.

Damiani, J. (2019, September 3). *A voice deepfake was used to scam a CEO out of $243,000.* www.forbes.com/sites/jessedamiani/2019/09/03/a-voice-deepfake-was-used-to-scam-a-ceo-out-of-243000/?sh=266449a22416

Fleming, S. (2019). Cognitive psychology. EDTECH. https://doi.org/10.1839474068.

Gopavaram, S. J., Grobler, M., Kim, D., Das, S., & Camp, L. J. (2021). Cross-national study on phishing resilience. *Proceedings of the Workshop on Usable Security and Privacy (USEC).* https://ssrn.com/abstract=3859057

Grobler, M., Gaire, R., & Nepal, S. (2021). User, usage and usability: Redefining human centric cyber security. *Frontiers in Big Data*, *4*. https://doi.org/10.3389/fdata.2021.583723

Hadnagy, C. (2010). Social engineering: The art of human hacking (P. Wilson, Foreword). Wiley. https://doi.org/10.1002/9781119433729.

Kuipers, S., & Schonheit, M. (2021). Data breaches and effective crisis communication: A comparative analysis of corporate reputational crises. *Corporate Reputation Review*, *25*(3), 176–197. https://doi.org/10.1057/s41299-021-00121-9

Lyngaas, S. (2023, December 5). *Hackers access profiles of nearly 7 million 23andMe customers.* https://edition.cnn.com/2023/12/05/tech/hackers-access-7-million-23andme-profiles/index.html

MacDonald, C. (2023, December 3). *23andMe hackers accessed ancestry information on millions of customers using a feature that matches relatives.* www.engadget.com/23andme-hackers-accessed-ancestry-information-from-thousands-of-customers-and-their-dna-relatives-205758731.html

National Institute of Standards and Technology. (2023, August 8). *NIST Cyber security White Paper (CSWP) NIST CSWP 29 ipd.* https://nvlpubs.nist.gov/nistpubs/CSWP/NIST.CSWP.29.ipd.pdf

Núñez, R., Allen, M., Gao, R., Miller Rigoli, C., Relaford-Doyle, J., & Semenuks, A. (2019). What happened to cognitive science? *Nature Human Behaviour*, *3*(8), 782–791. https://doi.org/10.1038/s41562-019-0626-2

O'Reilly, R. C., & Munakata, Y. (2000). Computational explorations in cognitive neuroscience. https://doi.org/10.7551/mitpress/2014.001.0001

Paganini, P. (2022, September 15). *Crooks are using lures related to her majesty queen Elizabeth II in phishing attacks.* https://securityaffairs.com/135764/cyber-crime/queen-elizabeth-ii-phishing.html

Revell, E. (2024, January 4). 23andMe blames users for data breach, citing recycled passwords. www.msn.com/en-us/news/other/23andme-blames-users-for-data-breach-citing-recycled-passwords/ar-AA1mqmsH

RiskOptics. (2022, January 27). Cyber security and natural disasters: How mother nature affects cyber risk. https://reciprocity.com/blog/cybersecurity-and-natural-disasters/

Sharples, S. (2022, October 31). 10 per cent of Optus customers leave after cyberattack. *News.com.au*. https://www.news.com.au/finance/business/other-industries/10-per-cent-of-optus-customers-leave-after-cyberattack/news-story/431a0661233a698eb3a6d2bb7c68562c

Taylor, J. (2022, September 29). Optus data breach: Everything we know so far about what happened. *The Guardian*. https://www.theguardian.com/business/2022/sep/29/optus-data-breach-everything-we-know-so-far-about-what-happened

Trend Micro. (2019, September 5). Unusual CEO fraud via deepfake audio steals US$243,000 from UK company. https://www.trendmicro.com/vinfo/us/security/news/cyber-attacks/unusual-ceo-fraud-via-deepfake-audio-steals-us-243-000-from-u-k-company

Uebelacker, S., & Quiel, S. (2014). The social engineering personality framework. In *2014 Workshop on Socio-Technical Aspects in Security and Trust*. https://doi.org/10.1109/stast.2014.12

Voce, I., & Morgan, A. (2023, June 27). Cybercrime in Australia 2023. https://doi.org/10.52922/sr77031

Wang, Z., Sun, L., & Zhu, H. (2020). Defining social engineering in cyber security. *IEEE Access*, *8*, 85094–85115. https://doi.org/10.1109/access.2020.2992807

Wang, Z., Zhu, H., & Sun, L. (2021). Social engineering in cyber security: Effect mechanisms, human vulnerabilities and attack methods. *IEEE Access*, *9*, 11895–11910. https://doi.org/10.1109/access.2021.3051633

Chapter 4

Cybersecurity in australian higher education curricula

The SFIA framework

Anna J. Griffin and Nicola F. Johnson

4.1 INTRODUCTION

Technology has developed exponentially over the past two decades, and the cybersecurity industry has struggled to keep up with the unprecedented rise in cyberattacks. Higher education institutions (HEIs) globally have increased their cybersecurity courses in response to the increased threats. Cybersecurity as a discipline in HEIs is a relatively new area and has been steadily developing. Part of the development process as a new discipline has been establishing processes and developing curriculum within HEIs that provide students with accreditation. This process is not unique to HEIs within Australia. Many countries, including the United States, the United Kingdom, and much of Europe, have established accreditation processes, all unique to each country depending on their needs and criteria. However, processes that were established in the infancy of a discipline should take time to evaluate the effectiveness of its initial procedures and perhaps consider, as a discipline, the best way a discipline might mature. This chapter explores the deployment of skills frameworks in Australian Universities within the cybersecurity discipline. In Australia, the Skills Framework for the Information Age (SFIA, 2021) is the primary framework that underpins professional accreditation from the Australian Computer Society (ACS). The ACS (2022) is the primary accreditation available to cybersecurity courses at HEIs, and as such, the cybersecurity curriculum must consider the SFIA framework in its development. Considering international literature that identifies alternative frameworks are used, this chapter explores the pros and cons of using SFIA within Australian HEIs when developing cybersecurity curriculum and discusses possible alternatives.

4.2 CURRICULUM DESIGN

Defining curriculum is difficult as the literature has more than 120 definitions. Two definitions of curriculum poignant to Higher Education (HE) are curriculum as a product and curriculum as a set of performance objectives

DOI: 10.1201/9781032664859-4

(Marsh, 1997). Curriculum as a product includes the documentation surrounding objectives, goals, content, and assessment, which is publicly available from any unit or course handbook on HEI websites. The curriculum, as a set of performance objectives, focuses on what students should attain, including specific skills or knowledge in the specialised area being covered in a unit. Both these definitions align with the HE sector. The focus of curriculum within this chapter is examining the curriculum as a product and as a set of performance objectives that can be seen through documentation – it is not considering the 'hidden' curriculum. The hidden curriculum is the teaching and learning that occurs within HEIs that is not the written curriculum but occurs due to the approaches of staff and the interactions that occur with students, even though they were not the intended learning outcomes (Marsh, 1997). The hidden curriculum is very different to the written curriculum and is not within the scope of this chapter.

Recently, the cybersecurity curriculum has become an area of development within HEI. Santos et al. (2017) succinctly stated that cybersecurity is the 'preservation of confidentiality, integrity and availability of information in Cyberspace' (p. 1). Cybersecurity is defined by a Joint Task Force on Computing Curricula as follows:

> [A] computing-based discipline involving technology, people, information, and processes to enable assured operations in the context of adversaries. It involves the creation, operation, analysis, and testing of secure computer systems. It is an interdisciplinary course of study, including aspects of law, policy, human factors, ethics, and risk management.
> (Bishop et al., 2017; Blaine et al., 2021; Parrish et al., 2018)

Within this explanation of cybersecurity, the discipline's complexities are highlighted. Historically, cybersecurity has included aspects of several disciplines, including information security, digital forensics, network security, information warfare, etcetera, spurred on through the disciplines' origins and can be traced in military, business, law, and computer science. With a complex history, it is no wonder that curriculum development may experience difficulties with so many facets to consider. Until recently, cybersecurity was often added to existing information security or computing science courses, identifying it as a standalone cybersecurity discipline. It is not surprising that within the literature, many of the challenges identified regarding curriculum design for cybersecurity courses are as follows: (1) there is little consensus (Blaine et al., 2021), (2) there is limited agreement on what should be known by cyber professionals (Bishop et al., 2017), (3) there is no global curriculum for cybersecurity (Asghar & Luxton-Reilly, 2020), and (4), there is a mismatch between the key stakeholders, including academia, industry, and government (Hajny et al., 2021), argued as one of the main reasons for the skills shortage.

Every HEI is presented with a challenge when it comes to curriculum design. Consideration must be given to course structure, the knowledge, skills, and abilities (KSAs) students are to be taught and assessed, and how to schedule and meet university protocols about course and unit delivery. Further consideration must include the effective implementation of policies regarding assessment, the award of a qualification, and the provision of learning experiences throughout a student's course enrolment. Parrish et al. (2018) outlined two cybersecurity course design approaches. The first approach is when cybersecurity units and content are integrated into an existing computing course (computer science or information technology, for example), and it is renamed to reflect the addition of cybersecurity. The other approach is to develop a new stand-alone course reflecting the cybersecurity discipline. They explained that it was common practice for the past 20 years to take the first option and adapt an already established course. However, they noted a movement towards creating new stand-alone courses, given cybersecurity's increasing maturity as a discipline.

There is limited literature on deciding what content to include within the cybersecurity curriculum. Jones et al. (2018) are among the few who provided empirical evidence as to what should be included in the cybersecurity curriculum. Their study interviewed cybersecurity employees who had one of four roles within the 'Protect and Defend' specialty area, as defined by the National Initiative for Cybersecurity Education (NICE) framework. Although the research group interviewed the participants, questions were completed on a survey using the Likert scale, leading to quantitative data analysis that analysed the KSAs most important to their job and those that were the most difficult to learn. From this research, Jones et al. (2018) provided clear guidance to educators of what should be included in the curriculum within the 'Protect and Defend' specialty area.

Bishop et al. (2017) argued that the cybersecurity education curriculum should be based on a computer science foundation, including core cybersecurity knowledge and skills based on the Cybersecurity Curricula (CSEC, 2017) guidelines. In a similar thought, Blaine et al. (2021) claimed undergraduate courses were developed in one or two ways: first, as a deep exploration of a single technical discipline (so computer-based, similar to Bishop et al., 2017), or second, as a course examining both technical and human elements within a broader curriculum. They contended that courses should specialise in areas in high demand by industry. Both Bishop et al. (2017) and Blaine et al. (2021) asserted that cybersecurity curriculum should have a solid common foundation. For example, within Data Security, the essentials include basic cryptography, digital forensics, end-to-end secure communications, data integrity and authentication, and information storage security.

4.3 FRAMEWORKS

The literature on cybersecurity education has a substantial reference to using frameworks and guidelines when developing curricula in HEIs. In a recent scoping review by the first author (Griffin, 2024), 14 of 20 papers referred to using frameworks when developing curriculum. The terms 'framework' and 'guidelines' were often used interchangeably, with little differences apparent. Within this chapter, the framework terminology will be used throughout to examine skills frameworks used as a reference within the cybersecurity industry and education.

Brown (2020) explained that a skills framework provides stakeholders with a common terminology that can be used within an industry when referring to skills or competencies required for a specific sector of an industry. They discuss how it can facilitate skills management and include 'technical' skills specific to the discipline or the 'soft' skills transferable between disciplines, for example, communication, teamwork, and leadership. Brown and Parr (2018) state that, in theory, a skills framework should provide the same base of specific skills from which all users can work. Brown and Parr (2018) explain that most skill frameworks have typical goals that they aim to facilitate:

- 'Assessment of current skill proficiency for individuals.
- Identification of target skill proficiency for individuals.
- Assessment of current skill portfolios in an organisation.
- Identification of skill gaps in an organisation.
- Design of accurate job and role descriptions.
- Design training and educational programmes by training/educational providers' (p. 2).

The focus of this chapter is the use of frameworks in the design of training and education curriculum and courses; however, it is also essential that HEIs consider how to identify and assess particular skills proficiencies for individual students. Therefore, an essential purpose and role of frameworks within cybersecurity education is to help curriculum designers understand what skills and knowledge the cybersecurity discipline, and in turn the graduates, need to develop when completing a cybersecurity course. Using frameworks helps academics define the topics and themes fundamental to course development (Hajny et al., 2021).

Skills frameworks are also utilised in organisations for both employers and employees. For employers, frameworks are helpful when assessing staff skill proficiency and planning future staff upskilling. Accessing a skills framework through a self-service application allows employees to manage their skill profiles and identify training needs. Staff can use it to demonstrate proficiency through their work experience (Brown, 2020). Some argue that a skills framework should also emphasise soft skills, as this is an area employers claim graduates are not prepared in when entering the workforce (Brown, 2020). It is important to note that although there is no doubt a skills framework has a significant role for organisations, this chapter aims to examine the use of skills frameworks within HEIs and the development of curriculum within the cybersecurity discipline.

Many benefits are identified in the literature regarding using frameworks in curriculum development (van Oorschot, 2021). Within the literature, most HEI academics use some form of framework to develop the curriculum (Griffin, 2024). They use these frameworks for their needs and capabilities because they have confidence in them, as they were developed by academics, governmental departments, and computing societies, and they help to define their curriculum (Hajny et al., 2021; Sobel et al., 2019). HEIs in cybersecurity use a framework for curriculum development because they provide rigour and outline the KSAs needed for a particular work task and because the KSAs can easily transform into learning outcomes. The frameworks are also believed to provide a global standard for countries developing their cybersecurity education (Knapp et al., 2017; Read et al., 2017).

One of the critical papers informing this chapter was co-authored by Hajny et al. (2021). Given that it is heavily relied upon in this chapter, we now provide an overview of its content. Hajny et al.'s (2021) seminal study undertook a comprehensive analysis of curriculum design recommendations alongside an analysis of the curriculum content of 89 existing courses (degrees) and mapped it to cybersecurity work role requirements. It then presented the SPARTA Cybersecurity Skills Framework (CSF) based on the structure of the NICE Framework, delineating SPARTA topics with NICE competencies and aligning them with KSAs. Their article is one of the few to extensively engage with curriculum mapping and address the need for industry and cybersecurity education to align better.

4.4 SKILLS FRAMEWORK FOR THE INFORMATION AGE (SFIA)

The Skills Framework for the Information Age (SFIA) was initially developed by the British Computer Society (BCS) alongside a variety of organisations from the United Kingdom and incrementally modified over the last few decades. Over 30 predominantly European organisations were involved in the development of SFIA, including the Department of Education and Employment and Microsoft (Brown, 2020; Bailey & Morrell, 2011). SFIA is a framework designed for Information Technology (IT) professionals and, according to Brown (2020), may be the most widely accepted IT skills framework globally for *IT as a discipline*.

The purpose of SFIA was to develop a high-level competency framework providing a standardised view of the skills needed by IT professionals (Bailey & Morrell, 2011). The intention was to develop an IT framework that used the best ideas from other frameworks and create a universal framework for common IT skills. It is meant to assist individuals and organisations in managing their skills (Brown, 2020). However, does using SFIA provide Australia HEIs with the best framework for cybersecurity education?

SFIA is currently up to version 8, which was released in 2021. SFIA is structured through professional skills, which are in six skills categories, each with its own subcategories:

- Strategy and architecture
- Change and transformation
- Development and implementation
- Delivery and operation
- People and skills
- Relationships and engagement

There are seven levels of responsibility, basically a grading system for the difficulty of each professional skill. At each level, the expectation is that a person demonstrates different skills associated with that professional skill and can demonstrate progression. The seven levels of responsibility range from level 1, which is the most basic, to the complex level 7. Not all professional skills will have levels one through seven; it depends on the skill (SFIA, 2021). Within each level of responsibility are five generic attributes: autonomy, influence, complexity, business skills, and knowledge (Brown, 2020).

To provide an example, Figure 4.1 outlines the descriptions of the professional skills of Security Operations. As can be seen, this professional skill can be attained through levels one to six and does not have a seventh level. The language used in the description is described as 'natural' and does not come from a standardised dictionary on IT.

Bailey and Morrell (2011) explain that using a professional industry framework in the curriculum helps address skill gaps early on and provides industry-wide recognition. However, it should be noted that they wrote about SFIA in 2011, which was before the establishment of other cybersecurity-specific frameworks.

More recently, Brown (2020) stated that SFIA does not provide universal certification criteria, which limits its portability and utility due to its lack of mapping between typical job roles. Also, Brown (2020) asserts that SFIA does not represent soft skills, for example, teamwork, leadership, and communication, that employers often refer to as paramount in any graduate. Brown and Parr (2018) explain that SFIA is counterintuitive with its lack of soft skills and levels of responsibility. They explain that a manager may

Level 1	• Performs simple security administration tasks. Maintains relevant records and documentation.
Level 2	• Receives and responds to routine requests for security support. Maintains records and advises relevant persons of actions taken. • Assists in investigating and resolving issues relating to access controls and security systems. • Documents incident and event information and produces incident, exception, and management reports.
Level 3	• Investigates minor security breaches following established procedures. • Assists users in defining their access rights and privileges. Performs non-standard operational security tasks. • Resolves security events and operational security issues.
Level 4	• Maintains operational security processes and checks that all requests for support are handled according to agreed procedures. • Provides advice on defining access rights and installing and operating elementary physical, procedural, and technical security controls. • Investigates security breaches following established procedures and recommends required actions. Provides support and checks that corrective actions are implemented.
Level 5	• Monitors the application and compliance of security operations procedures. • Reviews actual or potential security breaches and vulnerabilities and ensures they are promptly and thoroughly investigated. Recommends actions and appropriate control improvements. • Ensures that security records are accurate and complete and that requests for support are handled according to agreed procedures. • Contributes to creating and maintaining policy, standards, procedures, and documentation for security.
Level 6	• Develops policies, standards, processes, and guidelines for ensuring automated systems' physical and electronic security. • Ensures that the policy and standards for security operations are fit for purpose, current, and correctly implemented. • Reviews new business proposals and provides specialist advice on security issues and implications.

Figure 4.1 Example of SFIA Security Operations (SFIA, 2021).

have high leadership skills but a lack of technical expertise, which cannot be demonstrated in SFIA. They also question the benefits of including generic attributes, sometimes making the framework inflexible and limiting.

It should be noted that SFIA was not written with education in mind. It would be difficult to use the information in Figure 4.1 to develop a unit or course of work from this set of descriptions alone and would require some work to convert it into learning outcomes, KSAs, and assessments.

4.5 TWO EXAMPLES OF ACCREDITING BODIES

The accreditation body in the United States is the National Centres of Academic Excellence (CAE) in Cyber Defence. The National Security Agency and the Department of Homeland Security in the United States played a significant role in establishing educational standards for cybersecurity through the CAE accreditation programme, focusing on Cyber Defence and Cyber Operations. While not a framework, the accreditation programme has strict criteria for certification (Hajny et al., 2021). However, due to its prescriptive nature, only a limited number of higher education institutions have achieved accreditation. Despite the challenges with achieving accreditation, the CAE was discussed in 9 out of the 20 papers in the scoping review (Griffin, 2024). Conklin (2018) explains that CAE accreditation courses improve existing training materials, making them suitable for an educational setting by designing recommended knowledge units and topics to be included in the curriculum. The CAE courses aimed to provide structure and guidelines for cybersecurity education.

In Australia, the Australian Computer Society (ACS) established itself as a professional association providing professional accreditation of IT skills. Therefore, many universities that offer courses related to IT seek to align their courses to the requirements of ACS's professional accreditation. In a current review of the cybersecurity undergraduate courses offered by Australian HEIs, more than 75% stated in their course handbook that they offered professional accreditation through ACS.

The ACS has established its own Core Body of Knowledge (CBoK) for Information and Communication Technology (ICT) Professionals, which uses SFIA as a basis for their CBoK (CBoK, 2015). Van Oorscot (2021) explained that a *Body of Knowledge* (BoK) encompasses the 'entirety of facts, current beliefs or understandings, concepts, and practices of an academic domain or profession . . . commonly divided into knowledge areas (KAs)' (p. 83). The CBoK has three areas (ICT professional knowledge, ICT problem-solving, and ICT management). The ACS claims its accreditation process ensures ICT degrees are appropriate for people starting their careers in ICT. It also claims that one of its goals is to support the mobility of ICT graduates internationally and, as such, is an authority for skills assessments for immigration purposes (ACS, 2022). It should be noted that in Australia, there is no requirement for people to have a licence to practice; therefore, the accreditation process is not compulsory. However, for people to migrate to Australia as a skilled migrant, they need to have their skills accredited.

The ACS accreditation process covers all ICT professions – not just cybersecurity – and the documentation on the ACS website is generic. The ACS does specify that they have accreditation for cybersecurity courses; however, this information is not easily attainable from their website. The ACS uses SFIA as their main content for the CBoK, stating professionals need to demonstrate level 3 (undergraduate) or level 5 (postgraduate) from SFIA in their areas of specialisation, depending on the specific role of the cybersecurity professional. To obtain accreditation for cybersecurity degree courses, at least eight subjects (units) from CBoK or similar must be included.

It makes sense that the ACS would use an established framework to develop its CBoK for the accreditation process; however, is SFIA the most appropriate framework for cybersecurity education in Australia? Should HEIs offering cybersecurity courses continue to seek accreditation from ACS and a generic ICT accreditation process when alternative and widely used cybersecurity discipline-specific frameworks are available? Two decades ago, when cybersecurity was merely developing as a concept, using ICT or IT generic processes would have been appropriate. However, it could be argued that cybersecurity as its discipline should be represented accordingly.

4.6 OTHER FRAMEWORKS

In a recent scoping review, the first author (Griffin, 2024) identified that SFIA was only mentioned in 1 of 20 cybersecurity curriculum development papers. Therefore, it is essential to identify and elaborate on other frameworks utilised within cybersecurity education. A detailed summary of the NIST NICE framework and the CSCE 2017 curriculum guideline is provided in the following section. The two outlined below are considered the most transferable globally. Following that, but in less detail, are brief overviews of CyBOK UK (2021) (different from the ACS CBoK) and SPARTA from Europe.

4.6.1 NIST national initiative for cybersecurity education (NICE): the United States

The National Initiative for Cybersecurity Education (NICE) is led by the National Institute for Standards and Technology (NIST; Newhouse et al., 2017) in partnership with the private sector, academia, and governments, and since 2010 has produced an array of documents, including the Workforce Framework for Cybersecurity (NICE Framework) (Hajny et al., 2021; Santos et al., 2017). This popular framework is based on 52 cybersecurity roles divided into 7 job categories and includes 33 specialty areas. The individual KSAs are mapped to each of the work roles and, as such, clearly outline the requirements for professionals and educational standards (Hajny et al., 2021; Sobel et al., 2019; Knapp et al., 2017; Parrish et al., 2018). This emphasis on the workforce makes it distinctive from the other frameworks. The NICE framework strengthens employers' position to (a) enhance the workforce and (b) develop employee careers through developing the employees themselves (Brown & Parr, 2018). The NICE framework links the job market with work roles through an extensive spreadsheet highlighting the necessary KSAs. The NICE Framework defines professional requirements, sets educational standards, and proposes best practices (Hajny et al., 2021; Parrish et al., 2018). The language used to express the KSAs uses the industry and cyber workforce dictionary. The use of this dictionary has created a common terminology. In a recent scoping review on cybersecurity curriculum development, the NICE framework was the most referenced (Griffin, 2024).

Knowledge	
K0001	Knowledge of computer networking concepts, protocols, and security methods.
K0002	Knowledge of risk management processes (e.g., methods for assessing and mitigating risk).
K0332	Knowledge of network protocols such as TCP/IP, Dynamic Host Configuration, Domain Name System (DNS), and directory services.
K0334	Knowledge of network traffic analysis (tools, methodologies, processes).
Skills	
S0007	Skill in applying host/network access controls (e.g., access control list).
S0053	Skill in tuning sensors.
S0054	Skill in using incident handling methodologies.
S0077	Skill in securing network communications.
Abilities	
A0123	Ability to apply cybersecurity and privacy principles to organisational requirements (relevant to confidentiality, integrity, availability, authentication, non-repudiation).
Task	
T0042	Coordinate with cyber-defence analysts to manage and administer the updating of rules and signatures (e.g., intrusion detection/protection systems, antivirus, and content blacklists) for specialised cyber-defence applications.
T0261	Assist in identifying, prioritising, and coordinating the protection of critical cyber-defence infrastructure and key resources.
T0335	Build, install, configure, and test dedicated cyber-defence hardware.

Figure 4.2 Example of NIST NICE Framework (Newhouse et al., 2017).

If we compare the work role of the Cyber Defence Infrastructure Support Specialist in the NICE framework to our example in Figure 4.1, both are similar work roles; the difference in the details of the two frameworks is notable as NICE has 34 KSA and task statements for the role. Each statement can be used to help develop a curriculum and is broken down into manageable tasks that could be easily assessed. All of these are very helpful when developing a curriculum. In Figure 4.2, we have included just a few examples of the KSAs and tasks for this work role in NICE. The other noticeable aspect of NICE is that the tasks are written using cognitive verbs from the Revised Bloom's Taxonomy (Anderson et al., 2001), which indicates to educators the level of cognitive difficulty each task needs to be set at.

4.6.2 CSEC2017 for cyber and CC2020 for computer science

The Computing Curriculum 2020 project produced a report in 2020 by the Joint Task Force, an updated version of a previous report produced in 2005. This report summarises curricula and content for several computing disciplines, including cybersecurity. This task force included individuals who represented numerous organisations and were led by the Association for Computing Machinery (ACM) and the IEEE Computer Society (CC2020 Task Force,

2020). The individual curriculum on cybersecurity that this report refers to is the 2017 Joint Task Force on Cybersecurity Education (CSEC, 2017).

The purpose of the CSCE 2017 was to provide global curriculum guidelines and structure for undergraduate courses in cybersecurity (Hajny et al., 2021). This framework was written with HEI teaching and learning in mind, and while industry, academics, and government stakeholders were involved in its development, its purpose for education is evident. As such, it groups the knowledge areas and units thematically and outlines what cybersecurity students should know as a level of proficiency (Hajny et al., 2021). CSEC 2017 is also not too prescriptive, allowing flexibility for curriculum design developers by using a multidisciplinary lens whilst concentrating on a specific cybersecurity element (Bishop et al., 2017).

As illustrated in Figure 4.3, there are eight of the CSEC 2017 knowledge areas, and each knowledge area is further broken down into knowledge

Knowledge Area: System Security

The System Security knowledge area focuses on the security aspects of systems composed of components and connections that use software. Understanding the security of a system requires viewing it not only as a set of components and connections but also as a complete unit in and of itself. This requires a holistic view of the system. Together with the Component Security and Connection Security Knowledge Areas (KAs), the System Security KA addresses the security issues of connecting components and using them within larger systems.

Essentials

- Holistic approach	- Access control	- Testing
- Security policy	- Monitoring	- Documentation
- Authentication	- Recovery	

Knowledge Units	**Topics**	**Description/Curricular Guidance**
System Control		This knowledge unit examines the security considerations involved in controlling the system itself. It includes detecting, compensating for, defending against, and preventing attacks.
	Access control	This topic focuses on controlling access to resources and the integrity of the controls rather than their controlling access to data, which is covered in the Data Security knowledge.
	Authorisation models	This topic covers authorisation management across many systems and the distinction between authentication and authorisation.
	Intrusion detection	This topic covers anomaly, misuse (rule-based, signature-based), and specification-based techniques.
	Attacks	This topic covers attack models (such as attack trees and graphs) and specific attacks.
	Defences	This topic includes ASLR, IP hopping, and intrusion tolerance.
	Audit	This topic covers logging, log analysis, and its relationship to intrusion detection.
	Malware	This topic includes examples such as computer viruses, worms, ransomware, and other forms of malware.
	Vulnerabilities models	This topic includes examples such as RISOS and PA and enumerations such as CVE and CWE.
	Penetration testing	This topic covers the Flaw Hypothesis Methodology and other forms (ISSAF, OSSTMM, GISTA, PTES, etc.).

Figure 4.3 CSEC2017 Knowledge Area System Security Curricular Guidelines.

Test item	Essential concepts and associated learning areas to test
1	Describe what a penetration test is and why it is valuable.
2	Discuss how to document a test that reveals a vulnerability.
3	Discuss the importance of validating requirements.

Figure 4.4 CSEC2017 Essential Concept Testing for the Knowledge Area System Security.

units designed with educators in mind (Cybersecurity Curricula, 2017: Curriculum Guidelines for Post-Secondary Degree Programs in Cybersecurity, 2018). These are designed using Bloom's Revised Taxonomy (Anderson et al., 2001) using the understanding and applying levels. Students are expected to achieve the learning outcomes demonstrating proficiency in the essential concepts (Jones et al., 2018).

Essential concepts and associated learning areas for each knowledge area are shown in Figure 4.4, with corresponding learning outcomes (Cybersecurity Curricula, 2017: Curriculum Guidelines for Post-Secondary Degree Programs in Cybersecurity, 2018). This framework style is easy to use with academics, as it provides detailed learning outcomes that can be included within their curriculum and learning opportunities developed to cover the essential concepts that can be designed.

4.6.3 The cybersecurity body of knowledge (CyBOK) – UK

The Cybersecurity Body of Knowledge (CyBOK, 2021) stemmed from a project funded by the National Cybersecurity Programme at the University of Bristol in the UK (Rashid et al., 2019). The project aimed to define the knowledge base for cybersecurity as a discipline and establish a foundation for educational courses. The project recognised the fragmented and disjointed state of cybersecurity knowledge and sought to follow the example of mature sciences in progressing from secondary to undergraduate education (Hajny et al., 2021). The CyBOK project identified 19 knowledge areas (Kas) and grouped them into five broad categories. It would be beneficial for educators when developing a curriculum. Unlike other frameworks, particularly those outside the United Staes, it clearly outlines aspects of the curriculum for the non-technical elements, such as human, organisational, and regulatory aspects, as well as adversarial behaviours.

4.6.4 Cybersecurity skills framework (SPARTA) – Europe

The paper by Hajny et al. (2021), the Strategic Programs for Advanced Research and Technology in Europe (SPARTA) Framework, was presented. The SPARTA framework was developed to address the skills gap in the

European Union by providing a common language, knowledge, and skills management. The SPARTA framework was created through research summarising the global content covered by undergraduates and postgraduates. Hajny et al. (2021) used their research to suggest the best practice for the content that should be covered in a cybersecurity education course. It is important to note that although the authors developed their framework, its main building blocks were drawn from the NICE Framework. The challenge with this framework is that it was drawn from their research, which examined only the six top university courses within each country, according to the Times Higher Education (THE) ranking; however, this is somewhat arbitrary as there is no guarantee the universities included in the research offered the 'best' cybersecurity courses in the country.

From the convenience of our workstation in a bid to conduct a brief desktop review, it was easy to identify what frameworks and accrediting bodies are used in other countries to accredit cybersecurity courses, which can be seen in Table 4.1. The bold font points to general frameworks for corporate organisations and government agencies and is not a framework for HE curriculum development.

Table 4.1 Country, Accrediting Body, and Curriculum Framework (Random Sampling)

Country	*Accrediting Body/Bodies*	*Framework(s)*
The United Kingdom	National Cybersecurity Centre (NCSC)	CyBOK UK
The United States	National Centers of Academic Excellence (CAE) in Cyber Defense/Cybersecurity	National Institute of Standards and Technology (NIST) National Initiative for Cybersecurity Education (NICE); CSEC 2017
New Zealand	National Cybersecurity Centre (NCSC)	**NCSC Cybersecurity Framework (based on NIST)**
South Africa	Institute of Information Technology Professionals (IITPSA)	Higher Education Information Technology South Africa (HEITSA) Cybersecurity Framework
Ukraine	National Accreditation Agency of Ukraine (NAAU)	**Ukraine Standards of Higher Education 1701 'Information Security'**
Australia	Australian Computing Society	**SFIA and CBok; South Australian Cybersecurity Framework; Essential Eight**
Czech Republic	National Cyber and Information Security Agency; National Accreditation Bureau for Higher Education 'Security Sector Fields'	Czech National Qualifications Framework in Cybersecurity

The SFIA website (2021) explains that the framework is not designed to be specific to a discipline; it states that the scope is too broad and too specific. SFIA provides the universal framework for skills and competency levels, and then professional disciplines should develop formal bodies of knowledge. Why is Australia using SFIA when other countries and their accrediting bodies use higher education curriculum frameworks specifically developed for cybersecurity as a discipline?

4.7 DISCUSSION

When examining the purposes of the frameworks discussed, it becomes clear that SFIA was not written with the intention of education, as its primary purpose was for organisations and employers to manage IT skills, and it is not specific to cybersecurity. Although it does cover the theme of cybersecurity in its professional skills, it is not explicitly written with cybersecurity in mind, viewing it as a sub-discipline rather than its discipline.

The other frameworks, particularly NIST NICE and CSEC 2017, were written with cybersecurity higher education in mind after the early development stage. It is clear how academics can use them to develop curriculum for units within a degree. CSEC2017 is particularly for undergraduate curriculum in cybersecurity; this framework does not have organisations in mind, but rather, the students and their learning outcomes. This emphasis on learning outcomes can be seen in the structure of knowledge areas, units, and essential concepts. The language is in smaller chunks, easily converted into learning outcomes, providing clear topics and ideas for lessons and assessments.

The NICE framework is unique in that it is for organisations *and* education. The framework is developed around the different work roles within cybersecurity, and if HEIs decide to use it when curriculum planning, they could backward map from the work roles most in demand. Its structure sets out knowledge, skills, abilities, and tasks, which can all be used in the planning of teaching and learning within a course or individual units.

SFIA's descriptions at each level would require many adaptions to translate into appropriate learning outcomes for developing a cybersecurity curriculum. It lacks specific details about KSAs and tasks that these roles would demonstrate, and in turn, what students need to demonstrate to be proficient, that is, in terms of completing assignments or exams successfully. It should also be noted that the SFIA Foundation is now mapping itself to the NICE Framework, which demonstrates the global importance of the NICE Framework as it is recognised as the leading global framework in cybersecurity.

In the context of cybersecurity education in higher education, it can be inferred that frameworks have a crucial function in developing courses. The available research indicates specific frameworks may be more widely utilised by academics globally, with the National Institute of Standards and Technology (NIST) NICE Framework being a prominent example. The NICE framework has the potential to serve as a global reference point for work roles and the knowledge, abilities, and skills necessary for the cybersecurity industry. Its adaptable nature makes it a suitable choice for many higher education institutions, and its potential is further expanded with the recent addition of competencies by the NIST. This expansion makes it even more applicable to educational settings. In addition, the CyBOK UK framework (2021) may be more suitable to draw upon than SFIA as even the ACS CBOK V3.2 states that cybersecurity has a well-developed discipline body of knowledge in CyBOK UK. Thus, frameworks are essential tools for developing cybersecurity education in higher education, and certain frameworks hold more prominence in specific regions.

4.8 CONCLUSION

Two decades ago, when cybersecurity education was emerging, it was appropriate to use SFIA as the skills framework on which to base the HEI curriculum due to a lack of other adequate frameworks. Nevertheless, that is no longer the case as frameworks have been developed with cybersecurity-specific education at the forefront, demonstrating increasing confidence in the maturity of cybersecurity as a discipline. The ACS provides the only accreditation in Australia now, though it is not mandated for cybersecurity industry work requirements, and HEIs do not require it for their courses. It is clear from the discussion in this chapter that perhaps it is time for HEIs to question whether SFIA is the most appropriate framework for cybersecurity courses to be based on and to question the validity of ACS still using SFIA when there are alternative widely used comprehensive frameworks available.

REFERENCES

ACS. (2022). *Australian Computer Society*. www.acs.org.au/home.html

Anderson, L. W., Krathwohl, D. R., & Bloom, B. S. (2001). *A taxonomy for learning, teaching, and assessing: A revision of bloom's taxonomy of educational objectives* (Complete ed.). Longman.

Asghar, M. R., & Luxton-Reilly, A. (2020). A case study of a cyber security programme. *Proceedings of the 51st ACM Technical Symposium on Computer Science Education*. https://doi.org/10.1145/3328778.3366918

Bailey, S., & Morrell, C. (2011, September 5–9). *The implementation of skills framework for the information age (SFIA) at INACAP University Chile*. Invited Presentation presented to: Embedding Skills Framework for the Information Age (SFIA) in the Curriculum, INACAP Technological University of Chile. http://nectar.northampton.ac.uk/3933/

Bishop, M., Burley, D., Buck, S., Ekstrom, J. J., Futcher, L., Gibson, D., Hawthorne, E. K., Kaza, S., Levy, Y., Mattord, H., & Parrish, A. (2017). Cyber security curricular guidelines. In *IFIP advances in information and communication technology* (pp. 3–13). https://doi.org/10.1007/978-3-319-58553-6_1

Blaine, R. W., Blair, J. R. S., Chewar, C. M., Harrison, R., Raftery, J. J., & Sobiesk, E. (2021). Creating a multifarious cyber science major. *Proceedings of the 52nd ACM Technical Symposium on Computer Science Education*. https://doi.org/10.1145/3408877.3432462

Brown, J. (2020). An examination of the skills framework for the information age (SFIA) Version 7. *International Journal of Information Management, 51*, 102058. https://doi.org/10.1016/j.ijinfomgt.2019.102058

Brown, J., & Parr, A. (2018). ICT skills frameworks: Do they achieve their goals and users' expectations? *Advanced Journal of Professional Practice*, *1*(2), 38–47. https://doi.org/10.22024/UniKent/03/ajpp.506

CBoK. (2015). *The ACS core body of knowledge for ICT professionals CBoK*. www.acs.org.au/content/dam/acs/acs-skills/The-ACS-Core-Body-of-Knowledge-for-ICT-Professionals-CBOK.pdf

CC2020 Task Force. (2020). *Computing curricula 2020: Paradigms for global computing education*. https://doi.org/10.1145/3467967

Conklin, W. A. (2018). What constitutes core in cyber security curriculum. *Journal of the Colloquium for Information System Security Education*, *5*(2). https://cisse.info/journal/index.php/cisse/article/view/85/CISSE_v05_i02_p09.pdf

Cybersecurity Curricula (CSEC). (2017). *Curriculum guidelines for post-secondary degree programs in cyber security (2018)*. Association for Computing Machinery Joint Task Force on Cybersecurity Education. https://cybered.hosting.acm.org/wp/

CyBOK UK. (2021). *Knowledge base – CyBOK v1.1*. www.cybok.org/knowledgebase1_1/

Griffin, A. J. (2024). *Graduates' work readiness? Cyber security curriculum in Australian Universities* (Unpublished PhD thesis). Edith Cowan University. https://ro.ecu.edu.au/cgi/viewcontent.cgi?article=3323&context=ecuworks2022-2026

Hajny, J., Ricci, S., Piesarskas, E., Levillain, O., Galletta, L., & De Nicola, R. (2021). Framework, tools, and good practices for cyber security curricula. *IEEE Access*, *9*, 94723–94747. https://doi.org/10.1109/access.2021.3093952

Jones, K. S., Namin, A. S., & Armstrong, M. E. (2018). The core cyber-defense knowledge, skills and abilities that cyber security students should learn in school. *ACM Transactions on Computing Education*, *18*(3), 1–12. https://doi.org/10.1145/3152893

Knapp, K. J., Maurer, C., & Plachkinova, M. (2017). Maintaining a cyber security curriculum: Professional certifications as valuable guidance. *Journal of Information Systems Education*, *28*(2), 101–114. https://jise.org/Volume28/n2/JISEv28n2p101.html

Marsh, C. J. (1997). *Planning, management and ideology: Key concepts for understanding curriculum* 2. Routledge. https://doi.org/10.4324/9781315786698

Newhouse, W., Keith, S., Scribner, B., & Witte, G. (2017). *National initiative for cyber security education (NICE) cyber security workforce framework*. National Institute of Standards and Technology. https://doi.org/10.6028/nist.sp.800-181

Parrish, A., Impagliazzo, J., Raj, R. K., Santos, H., Asghar, M. R., Jøsang, A., Pereira, T., & Stavrou, E. (2018). Global perspectives on cyber security education for 2030: A case for a meta-discipline. *Proceedings Companion of the 23rd Annual ACM Conference on Innovation and Technology in Computer Science Education*. https://doi.org/10.1145/3293881.3295778

Rashid, A., Chivers, H., Danezis, G., Imperial, E. L., & Martin, A. (2019). *The cyber security body of knowledge*. www.cybok.org/media/downloads/cybok_version_1.0.pdf

Read, H., Sutherland, I., Xynos, K., Drange, T., & Sundt, E. (2017). The impact of changing technology on international cyber security curricula. In *Lecture notes in computer science* (pp. 518–528). https://doi.org/10.1007/978-3-319-58460-7_36

Santos, H., Pereira, T., & Mendes, I. (2017). *Challenges and reflections in designing cyber security curriculum*. In 2017 IEEE World Engineering Education Conference (EDUNINE). https://doi.org/10.1109/edunine.2017.7918179

SFIA. (2021). *Skills for the information age framework*. https://sfia-online.org/en

Sobel, A., Parrish, A., & Raj, R. K. (2019). Curricular foundations for cyber security. *Computer*, *52*(3), 14–17. https://doi.org/10.1109/mc.2019.2898240

Van Oorschot, P. C. (2021). Coevolution of security's body of knowledge and curricula. *IEEE Security & Privacy*, *19*(5), 83–89. https://doi.org/10.1109/msec.2021.3093735

Chapter 5

Dark echoes

The exploitative potential of generative AI in online harassment

Adrian Wood

5.1 BACKGROUND

5.1.1 Introduction to generative AI

Generative AI, a significant branch of artificial intelligence, represents a transformative leap in digital technology. Unlike traditional AI primarily focused on interpreting and learning from data, generative AI extends this concept to creating new, original content that closely mimics human-generated data (Goodfellow et al., 2020). The technology has become increasingly sophisticated, driven by advancements in neural networks and machine learning algorithms. These advancements enable generative AI to not only understand complex patterns in vast datasets but also to use this understanding to generate new data that bears a striking resemblance to the original input. This capability has opened up new possibilities across various domains, transforming how machines interact with data and creativity.

5.1.2 Applications across various domains

The versatility of generative AI is evident in its wide range of applications. In the realm of art and entertainment, it has revolutionised the creative process. Artists and designers are using AI algorithms to generate new visual art pieces, music compositions, and interactive media experiences, pushing the boundaries of traditional creativity (Zhu et al., 2017). Generative AI has been a boon for medical imaging and drug discovery in healthcare. It assists in creating detailed medical images for diagnosis and simulating drug interactions, thus speeding up research and development processes (Miotto et al., 2018). In finance, generative AI aids in complex tasks such as predicting market trends, assessing credit risk, and detecting fraudulent activities, thus enhancing accuracy and efficiency in financial decision-making (Cao et al., 2015; Mizuta, 2016; Vieira & Sehgal, 2017).

 DOI: 10.1201/9781032664859-5

5.1.3 Fundamentals of generative AI algorithms

The core of generative AI lies in its sophisticated algorithms, primarily based on deep learning methodologies such as Generative Adversarial Networks (GANs) and Variational Autoencoders (VAEs). GANs, for instance, involve two neural networks – a generator and a discriminator – that are trained simultaneously. The generator creates data, while the discriminator evaluates its authenticity. This adversarial process results in the generation of highly realistic data (Goodfellow et al., 2020). VAEs, on the other hand, focus on encoding data into a compressed representation and then reconstructing it, which is particularly useful in generating new data that retains crucial characteristics of the original set (Kingma & Welling, 2013).

5.1.4 Autonomous creation of lifelike content

One of the most remarkable aspects of generative AI is its ability to produce content virtually indistinguishable from authentic human-created material autonomously. This capability is limited to visual content and extends to text, audio, and video. For example, recent advancements have seen AI generating realistic human faces, writing coherent and contextually relevant articles, and even creating deepfake videos that are incredibly difficult to distinguish from actual footage (Radford et al., 2015). This ability of generative AI to mimic human creativity has profound implications for industries ranging from entertainment to journalism, redefining the boundaries between human- and machine-generated content. In exploring these advancements, DiPaola et al. (2018) discuss how AI generative techniques can be informed by cognitive theories of human creativity, thus bridging the gap between artificial and human creative processes.

However, the darker side of this technology is its potential for misuse, particularly in online harassment. The ability to create realistic synthetic content can be weaponised to fabricate convincing but false narratives, thus contributing to the spread of misinformation and manipulation (Chesney & Citron, 2018).

5.1.5 Ethical and security implications

While promising, the rapid development of generative AI brings forth significant ethical and security concerns. Its capability to create realistic content can be exploited for malicious purposes, such as spreading misinformation or creating realistic deepfakes that can be used to deceive and manipulate public opinion. These concerns are particularly alarming as deepfakes and AI-generated content become more sophisticated and harder to distinguish from authentic material. Partadiredja et al. (2020) highlight the socio-ethical implications of AI-generated media content, underscoring the risk of deepfakes contributing to misinformation and fake news proliferation. The ethical challenges in data-driven AI systems, as discussed by Henderson et al. (2018), underscore the need for careful consideration of the implications of AI in various domains, including generative AI.

5.1.6 Vulnerabilities exploited by malicious actors

The lifelike nature of content generated by AI systems presents a unique vulnerability that malicious actors exploit. There have been instances where generative AI has been used to create deepfake videos for political manipulation or to forge convincing textual content for phishing scams. The technology can also be used to create realistic images and narratives for online harassment, impersonating individuals and fabricating damaging information. Such misuse poses significant challenges to digital security and individual privacy, necessitating the development of robust detection and mitigation strategies (Chesney & Citron, 2018).

5.2 SIGNIFICANCE OF THE ISSUE

5.2.1 Escalation of online harassment through generative AI

The advent of generative AI has significantly altered the landscape of online harassment, introducing new dimensions and complexities to this pervasive issue. The technology's ability to create convincing, synthetic content has been co-opted by malicious actors, who use it to automate and amplify harassment campaigns. This misuse of AI blurs the boundaries between reality and fabrication, making it increasingly challenging to identify and combat such attacks. The enhanced realism and believability of AI-generated content can exacerbate the psychological impact on victims, intensifying the harm caused by online harassment (Chesney & Citron, 2018).

Online harassment, amplified by generative AI, has far-reaching consequences beyond individual victims. It threatens the integrity and inclusivity of digital communities, undermining the sense of safety and belonging essential for constructive online engagement (Citron, 2020). The pervasive nature of such harassment can lead to a chilling effect on free speech and expression, as users may self-censor to avoid becoming targets. This erosion of trust and safety within online spaces highlights the urgency of addressing the issue for individuals' well-being and online communities' health.

5.2.2 Legal and ethical quandaries

The rapid advancement of AI technologies, particularly those capable of generating content autonomously, challenges existing legal frameworks, which often lack the provisions necessary to address the intricacies of AI-generated content. This gap in legal preparedness becomes especially evident when considering the difficulties in tracing the origins of such content and in attributing responsibility, thereby complicating the pursuit of legal remedies for victims of AI-facilitated harassment.

From an ethical standpoint, using generative AI in harassment underscores profound concerns about accountability. This includes the responsibility of AI developers and platforms to safeguard against misuse of these technologies. The works of Partadiredja et al. (2020) and Liu et al. (2020) provide insightful perspectives on these challenges. Partadiredja et al. explore the socio-ethical implications of AI-generated media, shedding light on how these concerns manifest in online harassment scenarios. Meanwhile, Liu et al. offer an in-depth analysis of the legal disruptions caused by AI, highlighting the pressing need for legal systems to adapt and respond to the evolving AI landscape.

Further adding to this discourse, Henderson et al. (2018) emphasise the critical role of ethical considerations in AI development and governance. Their insights advocate for a proactive approach to ensuring accountability in AI usage, particularly in applications that might influence public discourse or individual rights.

5.2.3 Strategies for detection, prevention, and redress

The advent of generative AI marks a critical turning point in online harassment, highlighting the pressing significance of this evolving issue. The inherent capacity of generative AI to fabricate convincing synthetic content has become a tool for malicious actors, enabling them to orchestrate more sophisticated and damaging harassment campaigns. This misuse of AI technology represents a significant challenge, as it blurs the distinction between reality and artificiality, complicating identifying and mitigating such attacks.

In their comprehensive analysis, Flynn et al. (2021) detail the exploitation of AI for generating deepfakes, a form of digital impersonation that can be used for defamation or psychological harm. The legal and ethical implications of these AI-generated deepfakes accentuate the need for a robust response from both the legal system and law enforcement agencies.

Birks and Clare (2023) bring to light the impact of AI in academic environments, demonstrating how it can be utilised to facilitate academic misconduct. This extends the issue's significance into educational integrity, underscoring the importance of adapting prevention and detection frameworks to include AI's evolving role.

Montasari (2022) sheds light on the national security risks posed by AI-facilitated cyberattacks. Instances such as AI-enabled phishing and adaptive malware highlight the critical need for advanced security measures and ethical guidelines in AI applications to safeguard against these threats.

The examples provided by Flynn et al. (2021), Birks and Clare (2023), and Montasari (2022) not only illustrate the diverse ways in which AI technology can be misused but also emphasise the profound necessity of addressing AI-facilitated online harassment. The advancement of AI technology, with its growing sophistication and accessibility, escalates the potential for its exploitation in harmful ways. This advancement necessitates a proactive and comprehensive approach to safeguard individuals and communities from the detrimental effects of such harassment.

Developing integrated, multifaceted strategies is crucial in responding to these challenges effectively. Technologically, this involves creating advanced AI detection systems capable of distinguishing between authentic and AI-generated content and implementing robust cybersecurity measures to prevent AI-driven cyberattacks. Legal strategies require updating existing laws and regulations to encompass the new forms of abuse enabled by AI, ensuring that legal frameworks are equipped to handle cases of deepfake abuse, AI-facilitated academic misconduct, and other AI-driven cybercrimes.

Additionally, educational responses are vital in raising awareness and understanding of AI technology, its potential for misuse, and the ethical considerations surrounding it. This could involve integrating AI literacy into educational curricula, conducting public awareness campaigns, and training law enforcement and legal professionals on the nuances of AI-facilitated crimes.

Furthermore, there is a need for cross-sector collaboration involving policymakers, tech companies, educators, and legal experts to ensure a cohesive and effective response to the challenges posed by AI-facilitated online harassment. This collaboration is essential for staying ahead of the rapidly evolving technology and for developing strategies that are adaptable and responsive to future advancements in AI.

5.3 GENERATIVE AI AND ONLINE HARASSMENT: AN UNHOLY ALLIANCE

5.3.1 Exploitative potential

The advent of generative AI, particularly Generative Adversarial Networks (GANs), has ushered in an era of unprecedented lifelike content creation, blurring the lines between reality and synthetic fabrication. While this technological advancement holds immense potential, it also harbours a darker side, manifesting in the exploitative potential of AI in augmenting online harassment and disinformation campaigns. This emerging capability of AI to produce content that is virtually indistinguishable from authentic material raises profound concerns, especially as it becomes increasingly implicated in creating deepfake videos and generating fake news. The consequences of this technological abuse are far-reaching, impacting individuals through reputational damage and emotional distress and posing severe challenges for detection and response by law enforcement and online platforms.

5.3.2 Lifelike content creation

The realism generated by AI, mainly through GANs, raises significant concerns. These algorithms have become adept at creating images, videos, and text, often indistinguishable from authentic content. This capacity has been exploited in creating deepfake videos and generating fake news, where the distinction between natural and synthetic becomes increasingly blurred. The implications of this are profound, as victims of online harassment find themselves embroiled in false scenarios, leading to reputational damage and emotional distress (Paris & Donovan, 2019).

5.3.3 Automation of harassment tactics

The automation capabilities of generative AI have allowed harassers to scale their activities, amplifying the impact and reach of their attacks. Traditional harassment methods, such as personal attacks, spreading false information, or cyberstalking, can now be automated and executed with unprecedented efficiency and scale. This automation not only increases the volume of harassment but also makes it more difficult to trace and counter. For instance, AI-generated text can be used to automate the creation of harmful or abusive messages, while deepfakes can create false images or videos to damage reputations or blackmail individuals. The speed and scale at which this content can be disseminated pose significant challenges for detection and moderation for individuals and online platforms.

The ease with which AI can now replicate human-like content also means that harassment campaigns can be sustained over prolonged periods with minimal effort from the perpetrators. This makes the harassment more persistent and pervasive, exacerbating its psychological impact on victims.

5.3.4 Challenges for detection and response

The sophistication of generative AI complicates the detection of synthetic content. Traditional content moderation tools, which rely on identifying known patterns of abusive language or imagery, may be ineffective against AI-generated content that is constantly evolving and can be tailored to evade detection algorithms. Additionally, the realistic nature of this content can make it challenging for human moderators to discern its authenticity.

Law enforcement agencies and online platforms face considerable challenges in responding to AI-facilitated harassment. The anonymous and decentralised nature of the internet, combined with the difficulty of attributing AI-generated content to specific individuals, complicates efforts to hold harassers accountable. This situation is further complicated by the international scope of the internet, which can involve different jurisdictions with varying laws and regulations regarding online harassment and the misuse of AI.

5.4 NEFARIOUS USES OF GENERATIVE AI

5.4.1 Social media

Facebook recognises that deepfakes can be misused on social media in several harmful ways, primarily due to their ability to create convincingly altered videos that depict people saying or doing things they never did (Meta, 2020).

Firstly, deepfakes present a significant risk in spreading disinformation. Their ability to generate fake news or misleading narratives that are hard to differentiate from actual footage makes it easy to deceive viewers, potentially swaying public opinion. Additionally, deepfakes can severely damage the reputations of individuals, including public figures, politicians, and celebrities, by fabricating scenarios or statements that harm their personal and professional lives.

Another area of concern is the manipulation of political processes, such as elections. Deepfakes can be used to misrepresent politicians or public figures, thereby influencing voter perceptions and sentiments. This technology also contributes to social and political polarisation by creating divisive and controversial content, which can heighten societal tensions and conflicts. Moreover, deepfakes' deceptive nature can psychologically impact viewers, leading to a general mistrust in media and confusion about what is real. Finally, there is a risk of exploitation for financial gain, where deepfakes could be used in scams and fraudulent activities to deceive individuals.

5.4.2 CEO fraud case

A notable case of generative AI misuse occurred in 2019 involving a UK-based energy company. The CEO was tricked into transferring $243,000 by fraudsters who used AI to simulate the voice of his German parent company's CEO. The synthetic voice mimicked the German CEO's accent and mannerisms so convincingly that the UK CEO believed he was speaking to his boss (Stupp, 2019). This incident highlights the emerging risks of AI in corporate fraud and the urgent need for enhanced security measures to protect against such sophisticated attacks.

5.4.3 Virtual kidnapping scam

The advent of AI in cybercrime has led to a new wave of virtual kidnapping scams. In a notable case from April 2023, Jennifer DeStefano received a call with a ransom demand, backed by a distressing voice recording eerily similar to her daughter's voice, creating a believable kidnapping scenario (Gibson & Hagen, 2023).

Criminals initiate these scams by collecting voice biometrics from online sources like social media. Tools like VoiceLab process these biometrics to create convincing, deepfake voices. This technology enables attackers to convincingly impersonate victims, using the generated audio in extortion schemes (Gibson & Hagen, 2023).

Virtual kidnappers combine AI with social engineering to make their scams more effective. They target individuals, especially youths and public figures, whose online presence makes their biometrics more accessible. The scams exploit misinformation and emotional manipulation, causing not just financial harm but also significant emotional trauma to the victims (Gibson & Hagen, 2023).

AI tools, including ChatGPT, are employed to scale these scams. Attackers can efficiently target victims by processing large datasets of potential victims and integrating them with AI-generated voice and video data. This approach significantly increases the profitability and scalability of virtual kidnapping operations (Gibson & Hagen, 2023).

5.4.4 Revenge pornography

The advent of deepfake technology, a remarkable outcome of advancements in generative artificial intelligence, presents a paradoxical challenge. While it offers innovative possibilities in various fields, its misuse in creating non-consensual deepfake pornography is alarming. This digital abuse involves superimposing individuals' faces onto pornographic content without their consent, resulting in a disturbing form of image-based sexual abuse that raises serious ethical, psychological, and legal issues.

The emergence of deepfake technology in pornography dates back to around 2017. This technology leverages sophisticated AI algorithms, specifically, Generative Adversarial Networks (GANs), developed in 2014 by Ian Goodfellow and his colleagues (Goodfellow et al., 2020), to generate convincingly realistic videos. These AI models can seamlessly superimpose faces onto different bodies, initially targeting celebrities but quickly expanding to include private individuals. This trend is particularly worrying as it signifies a lowering of barriers to creating deepfakes, making this technology accessible to a broader audience with varying intentions (Hao, 2021; Rawle, 2023).

The psychological toll on victims of non-consensual deepfake pornography is substantial and far-reaching. A notable example Hao (2021) highlighted was Helen Mort, a UK poet who found her face illicitly superimposed onto explicit content. The images were taken from her social media profiles, leading to a profound sense of violation and powerlessness. This case underlines the ease with which personal images can be exploited in the digital era, transforming them into instruments of abuse.

In a similar vein, an Indian journalist became a target of a deepfake pornographic campaign following her report on a sensitive issue. This malicious act not only caused her immense personal distress but also had professional repercussions, demonstrating the potential of deepfake technology to inflict widespread damage to individuals' reputations and mental health (Rawle, 2023).

These incidents underscore a troubling pattern where deepfake pornography serves as a means for harassment and abuse, often with gender-specific implications. The simplicity of creating and disseminating such content heightens the risk of harm, making it a critical concern in digital ethics and legal discourse.

5.4.5 Manipulation in financial markets

The financial sector has seen AI-driven misinformation used to manipulate markets. For instance, on 22 May 2023, an AI-generated image of black smoke coming from a supposed government building near the Pentagon was spread online, stoking fear in investors and causing a drop in the S&P 500 by 0.29%. The image was quickly identified as fake, and the US Department of Defense confirmed no attack (McCarthy, 2023).

According to Bloomberg, it is possible this was the first instance of an AI-generated image influencing the market. The image first appeared on Facebook and quickly spread to Twitter. While the image was quickly identified as fake, organisations including RT and ZeroHedge reported on the false image (Sorkin et al., 2023).

5.4.6 AI-Generated phishing emails

The emergence of AI-generated phishing emails has significantly impacted the cybersecurity landscape. Using generative AI tools, notably those similar to ChatGPT, has substantially increased sophisticated phishing attacks.

WormGPT, developed by an anonymous creator known as 'last' or 'laste', exemplifies this trend. Introduced in March 2021 and later sold on hacker forums, WormGPT is designed to assist cybercriminals in creating tools for phishing and other cyber-related activities (Erzberger, 2023). It is available for a monthly fee of $200 or an annual subscription of $1,700, indicating its accessibility to a wide range of malicious users (Erzberger, 2023).

According to a CNBC report, there has been a 1,265% increase in malicious phishing emails since the fourth quarter of 2022, a surge closely aligned with the advent of generative AI tools like ChatGPT (Violino, 2023). This increase includes a 967% rise in credential phishing, suggesting a shift towards more targeted and sophisticated phishing efforts.

The 'State of Email Security in an AI-Powered World' report further delves into these concerns, highlighting the transformative impact of AI on email security. The report underscores the necessity for advanced detection methods and a deep understanding of the evolving nature of phishing attacks in the AI era (Darktrace, 2023).

5.4.7 Synthetic identity fraud

Synthetic identity fraud involves using AI to create entirely new identities, combining natural and fake personal information to bypass security checks. This advanced form of identity fraud exploits the capabilities of generative AI, which can analyse vast datasets to synthesise credible but entirely fictional personal identities.

Generative AI learns from a large corpus of real-world identity data, including names, social security numbers, addresses, and other personal details. It then uses this data to fabricate new identities, often indistinguishable from authentic ones. These synthetic identities can include credible-sounding names, plausible addresses, and social security numbers that pass basic validation checks. However, unlike traditional identity theft, synthetic identity fraud does not directly steal a natural person's identity; instead, it creates an entirely new identity that can be harder to trace and identify as fraudulent.

This fraud poses significant challenges to financial institutions, credit agencies, and security systems. AI-generated synthetic identities can be used to apply for credit, open bank accounts, or engage in other financial activities, often leaving a trail of financial losses and credit issues. Moreover, these activities can remain undetected for extended periods, as no real victim can report the fraud.

One of the primary challenges in combating synthetic identity fraud is the sophistication of the AI algorithms used. These algorithms are constantly evolving, becoming more adept at mimicking real-life patterns of personal information. For instance, Wakefield Research (2021) highlights the increasing prevalence of this type of fraud in the digital age, emphasising the need for more advanced detection methods to keep pace with the evolving techniques of fraudsters.

5.4.8 Fabricated evidence in legal cases

In the dynamic realm of artificial intelligence (AI), the legal sector is confronting significant challenges due to the advent of fabricated digital evidence. Grossman et al. (2023) discuss the difficulty faced by legal systems in distinguishing between authentic and AI-generated evidence, such as documents, videos, and audio recordings, especially with the rapid advancement of GenAI technologies. They note that these developments pose substantial challenges in determining the authenticity of evidence, leading to increased litigation costs and complicating the judicial process (Grossman et al., 2023).

Grossman et al. (2023) emphasise the risks associated with AI tools capable of producing highly convincing deepfakes. These technologies, including notable examples like OpenAI's ChatGPT and Google's LaMDA, have advanced language processing capabilities, enabling them to manipulate original recordings and create new, misleading narratives. The concern is that defendants might struggle to refute profound fake evidence unless they can afford expensive expert analysis, posing a threat to justice (Grossman et al., 2023).

Despite AI's potential benefits in fields like healthcare, Grossman et al. (2023) urge caution against its unchecked deployment. They argue for a balanced approach, advocating for regulations to safeguard against misuse while avoiding outright bans. The EU Parliament's (2023) draft legislation on facial recognition technology is an example of comprehensive regulation. Nonetheless, Grossman et al. (2023) acknowledge AI regulation challenges, such as the lack of public sector expertise and the risk of hindering innovation.

5.4.9 Examination of tactics employed by harassers

The exploitation of generative AI by harassers demonstrates their ability to blend traditional harassment tactics with advanced technology. The use of deepfake videos in political manipulation, the creation of non-consensual pornography, and the fabrication of legal evidence show a disturbing trend of using AI to harm individuals and institutions. These tactics often rely on the shock value and believability of AI-generated content, exploiting trust and gaps in legal and technological defences. Understanding and anticipating these tactics is crucial in developing comprehensive strategies to counter this new form of digital harassment.

Examining these case studies reveals the multifaceted nature of online harassment augmented by generative AI. It underscores the critical need for interdisciplinary efforts involving legal, technological, and educational measures to address the evolving landscape of digital harassment and ensure the responsible use of AI technologies.

5.5 LEGAL AND ETHICAL QUANDARIES

The advent of artificial intelligence (AI) heralds transformative changes across societal, ethical, and legal dimensions, challenging existing paradigms and prompting a re-evaluation of traditional frameworks. The rapid evolution of generative AI technologies, exemplified by large generative AI models (LGAIMs), brings to the forefront the complexities of AI-driven innovation and its implications. These technologies, capable of generating synthetic content indistinguishable from actual human output, pose unique challenges, especially in online harassment, copyright issues, and the broader societal impacts of AI (Hacker et al., 2023).

Generative AI's ability to blur the lines between reality and fiction introduces significant legal hurdles, particularly in addressing AI-driven harassment.

Traditional legal frameworks designed for more straightforward forms of abuse find it challenging to tackle the nuances and complexities introduced by generative AI. This discrepancy stems from the technology's capability to produce lifelike content, complicating the legal identification and prosecution of harassment. Furthermore, the scalability and anonymity afforded by AI technologies enable perpetrators to conduct extensive harassment campaigns efficiently and often undetectably, raising critical questions about the adequacy of existing laws and the need for adaptive, forward-looking legal approaches (Hacker et al., 2023).

As we delve into the ensuing sections, we will examine the ethical, societal, and legal implications of AI-generated content, the necessity for robust legal and regulatory frameworks tailored to the realities of generative AI, and the overarching goal of fostering a 'Good Society' in the age of AI. Through this analysis, we aim to underscore the imperative for collaborative efforts among lawmakers, technologists, and society to navigate the complexities of AI governance and ensure that AI serves the public good while mitigating potential risks and harms.

5.5.1 Legal challenges in addressing generative AI-driven harassment

The rapid advancement of generative AI significantly challenges existing legal frameworks to address online harassment. Traditional laws, designed for simpler abuse forms, struggle to tackle the complexities of AI-driven harassment. This discrepancy stems from generative AI's ability to create lifelike content that blurs reality and fiction, making it difficult to define and prosecute harassment legally. Additionally, the anonymity and scalability provided by AI technologies allow perpetrators to conduct extensive harassment campaigns efficiently and often undetectably.

The sophistication of AI-generated content complicates the identification of authentic versus synthetic media, challenging victims, legal practitioners, and policymakers alike. Existing legal statutes, reliant on clear definitions of abuse, fall short of addressing the nuanced issues presented by AI. The international nature of the internet further complicates the jurisdictional enforcement of laws against AI-enabled harassment.

Addressing these challenges requires re-evaluating and potentially overhauling current legal approaches to online harassment and adapting them to the rapid development of AI technologies. Legal systems need flexible, technology-agnostic definitions of harassment and international cooperation to tackle AI-facilitated abuse effectively. Collaborative efforts among lawmakers, legal scholars, and technologists are crucial to developing protective measures against AI-enabled abuse while promoting responsible AI use.

5.6 AI IN CREATIVE INDUSTRIES AND COPYRIGHT CHALLENGES

Integrating artificial intelligence (AI) into creative industries has sparked a significant re-evaluation of copyright laws, particularly as generative AI technologies advance. These technologies can produce creative content, including music, literature, and visual arts, challenging traditional notions of authorship and ownership. The case of Stable Diffusion, where AI-generated images led to copyright infringement lawsuits, exemplifies the growing tensions between AI innovation and copyright protections (Shroff, 2024).

Generative AI systems, such as GPT models used in text generation and Stable Diffusion for image creation, rely on massive datasets, often containing copyrighted materials, to train their algorithms. This raises critical questions about the rights of original creators whose works contribute, without explicit permission, to training these AI systems (Shroff, 2024). The legal quandary extends to determining ownership of AI-generated content, applying 'fair use' principles, and compensation for creators of source materials.

The rapid development of AI technologies introduces unique forms of creative expression not fully anticipated by existing legal frameworks. Hacker et al. (2023) note that the legal landscape surrounding generative AI and copyright law is ambiguous, necessitating a nuanced understanding of the AI value chain and targeted regulatory approaches. The complexity of AI-generated content complicates the identification of original versus synthetic media, challenging legal practitioners and policymakers to adapt traditional copyright statutes to the digital age's realities.

Moreover, the international nature of the internet further complicates the jurisdictional enforcement of copyright laws against AI-enabled content creation. Addressing these challenges requires a collaborative effort among lawmakers, legal scholars, and technologists to develop protective measures that encourage responsible AI deployment and use while ensuring creators are duly recognised and compensated.

5.6.1 Insufficiency of legislation in the face of AI advancements

The rapid advancement of generative AI technologies has created significant challenges for existing legal frameworks. Current legislation, primarily developed before the widespread adoption of sophisticated AI systems, struggles to address the complex issues arising from these technologies. One of the critical areas affected is intellectual property law. As generative AI continues to evolve, legal and ethical questions regarding the impact on artists' rights have become more prevalent (Poland, 2023). These issues highlight the need for updated laws that adequately address the nuances of AI-generated content and its implications.

Moreover, the legal lexicon is expanding to include new terms such as AI hallucination, AI bias, and generative AI, which reflect the emerging complexities in the field (Sall, 2024). This evolution in terminology indicates a shift in the legal landscape, necessitating a re-evaluation of existing laws to encompass the broad implications of AI technologies. The development of generative AI has outpaced the current legal structures, leading to a gap in regulation and oversight. This gap presents challenges in adequately protecting individuals and entities from potential abuses or infringements facilitated by AI technologies.

Additionally, the impact of evolving AI models on the legal sector is profound. Generative AI tools, ranging from language models like GPT technologies to art generation applications, reshape these technologies' legal considerations and ethical boundaries (Legaltech News, 2024). The legal community is now tasked with understanding and staying abreast of these advancements to ensure that laws remain relevant and effective in safeguarding rights and maintaining ethical standards in the digital age.

The insufficiency of current legislation in the face of rapid AI advancements is evident. It underscores the urgency for lawmakers and legal professionals to adapt and develop new regulatory frameworks to address the unique challenges of generative AI technologies.

5.6.2 Legal and regulatory frameworks for AI

The legal and regulatory landscapes for artificial intelligence (AI) are evolving in response to the rapid advancement of AI technologies and their profound societal implications. These frameworks are crucial for addressing ethical, privacy, security, and accountability issues, ensuring AI's development and deployment align with societal values and human rights.

Global Perspectives on AI Regulation: The European Union has been at the forefront of developing comprehensive AI regulation, notably through the proposed AI Act. This legislative framework categorises AI systems based on their risk levels, from unacceptable to minimal risk, imposing specific obligations on AI developers and deployers accordingly. Before deployment, high-risk AI applications, such as those involving biometric identification or critical infrastructure management, are subject to stringent requirements, including transparency, data governance, and human oversight (European Commission, 2021). Similarly, the General Data Protection Regulation (GDPR) sets a high data protection and privacy standard, impacting AI systems processing personal data by mandating user consent, data minimisation, and the right to explain automated decisions.

In contrast, the United States adopts a more sector-specific approach, focusing on guidelines and policies rather than a comprehensive national framework.

Initiatives like the National AI Research and Development Strategic Plan and executive orders on AI outline principles for fostering innovation while ensuring ethical, secure, and trustworthy AI development. This approach reflects the dynamic nature of AI technology and its diverse applications across different sectors (National Science and Technology Council, 2016).

Challenges in AI Regulation: One of the primary challenges in regulating AI is the technology's rapid evolution, which can outpace legislative processes. Legal uncertainty arises when existing laws struggle to accommodate new AI capabilities, creating a need for legal frameworks that are both flexible and technology agnostic. This uncertainty is compounded by AI's dual status as agent and object, challenging traditional legal categories and necessitating clarification or refinement of laws concerning responsibility and liability (Liu et al., 2020).

Moreover, international collaboration is essential for harmonising AI regulations, given the global nature of AI development and deployment. Efforts by international organisations, such as the OECD's AI principles, aim to provide a foundation for ethical AI governance, emphasising respect for human rights, democratic values, and diversity. These principles guide national policies and foster international cooperation in addressing AI's global challenges.

Future Directions: As AI continues to transform societies, the development of legal and regulatory frameworks must be anticipatory, adaptable, and inclusive. Engaging a broad range of stakeholders, including policymakers, technologists, civil society, and the public, is crucial for crafting regulations that balance innovation with ethical considerations and social welfare. The ongoing dialogue on AI regulation reflects a collective effort to steer AI development towards beneficial outcomes for humanity, emphasising the need for ethical AI development, transparency, and accountability.

5.6.3 Balancing free expression and user protection

The evolution of generative AI technologies brings the challenge of balancing the right to free expression with the need to protect individuals from AI-driven harassment. Current legal frameworks, developed in a pre-AI era, grapple with the complexities introduced by AI's creative capabilities. De Vries (2022) delves into this predicament by examining the status of AI-generated speech under Article 10 of the European Convention on Human Rights (ECHR). Unlike traditional expressive tools, AI's role in content creation blurs the lines of speech attribution, raising the question of whether machine-generated speech falls under the purview of freedom of expression rights.

This challenge is further complicated by the emergence of new types of AI-generated content, such as synthetic media and deepfakes, which create normative grey areas in legal discourse. De Vries discusses how such content can skirt established legal constraints, including copyright, data protection, and privacy. For instance, the use of AI in generating virtual child pornography or in propagating fake news poses significant ethical dilemmas. These instances highlight the difficulty distinguishing between protected and unprotected speech, a crucial distinction in a legal context.

Legal reforms that can keep pace with AI's advancements are critical in this rapidly evolving landscape. Policymakers, legal experts, and technologists must collaborate to develop frameworks that allow the free expression of AI-generated speech and ensure it does not infringe on individual rights.

5.7 ETHICAL CONCERNS

The emergence of AI-generated content, notably through generative AI technologies like GPT models and deep learning systems, has significantly blurred the lines between human- and machine-generated content, raising profound ethical and societal implications. The ability of these technologies to produce highly realistic text, images, and videos not only challenges our understanding of creativity and authorship but also poses risks related to misinformation, privacy, and the erosion of trust in digital media (Partadiredja et al., 2020).

One of the primary ethical concerns revolves around the potential for AI-generated deepfakes and synthetic media to propagate misinformation, affecting public opinion and democratic processes. The realistic nature of these creations can lead to confusion and difficulty distinguishing between authentic and manipulated contents, undermining the integrity of information and contributing to a post-truth society. This phenomenon necessitates a critical examination of the responsibilities of AI developers and platforms in mitigating the spread of harmful content while also considering the implications for freedom of expression and innovation (Qian et al., 2024).

Furthermore, using AI to create content that mimics human attributes raises significant privacy concerns. Generative AI's capability to replicate individuals' likenesses without consent introduces new forms of harassment and exploitation, highlighting the need for legal and ethical frameworks that protect individuals' rights in the digital age. Developing AI governance mechanisms that ensure transparency, accountability, and ethical use of AI technologies is crucial in addressing these challenges.

Additionally, the societal impact of AI-generated content extends to the labour market and the future of work. As AI technologies become more capable of performing tasks traditionally reserved for humans, including creative endeavours, there is a growing concern about job displacement and the creation of a 'useless class', as predicted by some scholars (Qian et al., 2024). This shift necessitates rethinking economic models, social systems, and education to prepare for a future where human and machine labour coexist.

A multi-stakeholder approach involving policymakers, technologists, and the public is essential to navigate AI-generated content's ethical and societal implications. Developing ethical guidelines, regulatory measures, and educational initiatives that promote responsible AI use while fostering innovation and protecting societal values is imperative. Collaborative efforts to enhance AI literacy and critical thinking among the general public can empower individuals to critically engage with AI-generated content, fostering a more informed and resilient digital society.

Invasion of Personal Space through AI-Generated Content: Generative AI enables content creation that can mimic an individual's appearance, voice, or personal attributes with alarming accuracy. Such technology can generate false representations or deepfakes, leading to serious privacy intrusions. As Guleria et al. (2023) discuss, the forensic, legal, and ethical issues surrounding AI applications like ChatGPT underscore the potential for these technologies to invade privacy and cause reputational harm.

Implications for Reputation and Psychological Well-Being: The misuse of generative AI in fabricating content can have dire consequences for individuals' reputations. False narratives or misleading representations created using AI can be spread rapidly online, damaging personal and professional reputations. The psychological impact of such attacks can be profound, leading to distress and trauma for victims. Elgesem (2023) highlights the ethical challenges of generative AI models under the AI Act, particularly concerning protecting personal privacy and preventing misuse.

Challenges in Legal Recourse and Redress: Victims of AI-driven privacy violations often face challenges in seeking legal recourse due to the novelty and complexity of the technology involved. Current privacy laws may not fully address the unique aspects of AI-generated content, leaving victims without adequate legal protection. Ferrara (2023) emphasises the need for legal systems to evolve and adapt to the advancements in AI to provide effective remedies for those affected by such privacy violations.

Ethical Responsibility of AI Developers: Developers and researchers involved in generative AI bear a significant ethical responsibility to consider the privacy implications of their creations. This includes implementing measures to prevent the misuse of AI technologies for violating personal privacy. As Meskys et al. (2020) discussed, regulating deepfakes and other AI-generated content requires a balance between innovation and ethical considerations, ensuring the responsible use of AI in preserving privacy.

Public Awareness and Digital Literacy: Raising public awareness about generative AI's capabilities and potential misuse is essential in mitigating privacy risks. Educating individuals about recognising and responding to AI-generated content can empower them to protect their privacy and seek appropriate recourse. Kiritchenko et al. (2021) underscore the importance of education and awareness in the ethical and human rights context of abusive language online, which can be extended to understanding privacy risks in AI.

Integrating generative AI into online harassment amplifies privacy concerns, necessitating a nuanced approach to address these challenges. A multifaceted strategy involving legal reform, ethical guidelines, public education, and responsible AI development is crucial in safeguarding individual privacy in the digital era. This approach not only aligns with legal imperatives but also upholds a commitment to respecting the rights and dignity of all individuals in the face of rapidly advancing AI technologies.

5.8 AI GOVERNANCE, STRATEGY, AND THE 'GOOD SOCIETY'

The quest for a 'Good AI Society' emphasises the imperative to harmonise the rapid advancements in artificial intelligence (AI) with the foundational principles of human dignity, equity, and societal well-being. AI governance and strategy are pivotal to navigating AI's ethical, legal, and social challenges, ensuring its deployment enhances rather than undermines the public good. This pursuit requires a comprehensive and forward-looking approach, integrating insights from blockchain technology to create a robust framework for ethical AI development and use.

Governance and Ethical Frameworks: The governance of AI involves the establishment of ethical frameworks, policies, and regulations that guide AI development towards positive societal impacts. Notable efforts by the European Union, the United States, and other global actors reflect a concerted move towards developing AI governance structures that balance innovation with ethical considerations. For instance, the proposed EU AI Act categorises AI systems according to risk levels, setting a precedent for future regulatory efforts focused on safety, transparency, and accountability (European Union, 2021).

Strategic Development of AI: AI development aligns AI innovations with societal goals and ethical standards. This includes fostering AI research prioritising human well-being, ensuring equitable access to AI benefits, and addressing potential harms. The strategic integration of blockchain technology can enhance AI's ethical governance by providing transparent, secure, and immutable records of AI operations, training data, and decision-making processes, thereby bolstering accountability and trust in AI systems.

Blockchain's Role in Ethical AI: Blockchain technology offers a decentralised framework supporting ethical AI governance by ensuring data integrity, transparency, and traceability (Ramos & Ellul, 2024). For example, blockchain can be used to document the provenance of training datasets, mitigating concerns over data bias and copyright issues in AI-generated content. Additionally, smart contracts on blockchain networks can enforce compliance with ethical guidelines and regulatory standards automatically, providing a novel mechanism for ethical AI governance.

Building a 'Good AI Society': Creating a 'Good AI Society' requires a multistakeholder approach that includes policymakers, technologists, civil society, and the public in the governance process. This approach should foster international collaboration to harmonise AI governance frameworks, ensuring they are adaptable to the evolving landscape of AI technologies and responsive to diverse societal needs. The integration of blockchain into AI governance structures could further democratise the development and use of AI, enabling more participatory and transparent governance models, as the blockchain's role in AI governance democratises AI development and use, embodying the principles of transparency and accountability essential for ethical AI ecosystems (Ramos & Ellul, 2024).

The path to a 'Good AI Society' is complex and multifaceted, demanding concerted efforts to align AI development with ethical principles, societal values, and human dignity. Effective AI governance and strategic planning are essential to navigate the challenges and opportunities presented by AI, with blockchain technology offering promising tools for enhancing the transparency, accountability, and ethical integrity of AI systems. By fostering collaborative, inclusive, and forward-looking approaches to AI governance and strategy, society can harness the transformative potential of AI to build a more equitable, sustainable, and flourishing future for all.

5.9 STRATEGIES FOR DETECTION AND PREVENTION

The emergence of generative artificial intelligence (AI) as a tool for creating sophisticated synthetic content has presented unprecedented challenges in distinguishing between authentic and manipulated digital media. This evolution has significant implications, mainly when such technology is used for nefarious purposes, including online harassment, misinformation campaigns, and digital fraud. Developing and implementing effective strategies for detecting and preventing AI-generated content misuse is imperative to address these challenges. These strategies encompass a multifaceted approach that combines the latest advancements in AI detection technologies, collaborative efforts across various industry sectors, and comprehensive public education and awareness initiatives. Additionally, they involve ongoing research and development, legal and policy measures to regulate AI use, and ethical considerations in AI development. Together, these strategies form a robust framework for safeguarding the digital landscape against the malicious use of AI, ensuring that technological advancements contribute positively to society rather than undermine trust and integrity online.

5.9.1 Advanced AI detection technologies

Researchers and technologists are focusing on several key strategies to combat the challenges posed by AI-generated content, particularly in identifying and mitigating the spread of deepfakes. According to Wang et al. (2020), the development of sophisticated AI detectors is paramount. These detectors employ advanced machine learning models, such as deep neural networks, trained to recognise specific patterns or anomalies unique to synthetic media. It is crucial to regularly update the algorithms to ensure these models remain effective against the rapidly advancing capabilities of generative AI technologies.

Zhou et al. (2021) emphasise the importance of continuously refining and updating detection tools to keep pace with the latest developments in AI-generated content. This process involves integrating new datasets and learning from real-world instances of AI misuse to enhance the accuracy and efficiency of detection systems.

Furthermore, Nguyen et al. (2022) highlight the use of digital watermarking as a critical technique in verifying the authenticity of digital media. Digital watermarking involves embedding invisible marks within digital content, which can be used to authenticate media and trace its origins. In an era increasingly dominated by synthetic media, such techniques play a vital role in distinguishing between genuine and manipulated content, thus helping maintain digital information's integrity.

5.9.2 Collaborative industry efforts

Russell et al. (2015) stress the importance of cross-platform collaboration as a pivotal measure in countering the misuse of AI-generated content. Encouraging cooperation among various platforms and technology companies is essential for sharing insights, detection tools, and best practices. Such collaborative efforts can significantly boost the collective capacity to identify and mitigate the adverse effects of AI-generated content, ensuring a more unified and effective response across the digital ecosystem.

Furthermore, establishing information-sharing networks is crucial for a timely and coordinated approach to new AI-generated threats. By sharing intelligence on emerging dangers, stakeholders can respond more swiftly and efficiently, staying one step ahead of potential threats. This collaborative approach underscores the necessity of open communication channels and cooperation among different entities involved in digital content management and security.

Dignum (2019) highlights the critical role of public–private partnerships in tackling the legal and regulatory challenges of AI misuse. Collaborations between governments, law enforcement agencies, and the private sector are indispensable for developing comprehensive strategies to combat AI-driven online abuse. These partnerships facilitate the exchange of crucial information and resources and contribute to formulating more effective and informed policies and regulations to govern the use and dissemination of AI-generated content, ensuring a safer and more secure digital environment for all users.

5.9.3 Public awareness and education

Wardle and Derakhshan (2017) emphasise the significance of educational campaigns in raising public awareness about the capabilities and associated risks of AI-generated content. These campaigns are crucial in informing users about recognising and reporting synthetic media, enhancing their ability to navigate the digital landscape safely and responsibly. In addition to awareness campaigns, integrating digital literacy into educational curricula is vital for equipping individuals with the skills to understand and critically evaluate online content. Prioritising the teaching of strategies to identify AI-generated content can empower people to discern between authentic and manipulated media, fostering a more informed and cautious online community.

Livingstone and Third (2017) advocate for the importance of media literacy programmes in helping individuals discern the credibility of online information. In an era increasingly characterised by the prevalence of synthetic media, such programmes are essential in empowering users to critically assess the authenticity and reliability of digital content, enabling them to make more informed decisions and engage with media more responsibly.

Wang's (2021) thesis, 'Generative Chatbot Framework for Cyber Grooming Prevention', presents a proactive approach to education and awareness regarding AI cyber grooming. The thesis introduces SERI, a sophisticated chatbot framework that employs deep reinforcement learning to simulate interactive dialogues between potential perpetrators and victims. This innovative educational tool is designed to enhance the understanding and recognition of cyber grooming behaviours, providing young internet users with essential knowledge and skills to identify and counteract such threats in digital environments. Through such proactive educational measures, individuals can be better equipped to navigate the complexities of the digital world, ensuring their safety and well-being online.

5.10 RESEARCH AND DEVELOPMENT

5.10.1 Investment in AI ethics research

AI ethics research plays a pivotal role in steering the development and utilisation of artificial intelligence technologies towards a path that aligns with social responsibility and ethical standards. As AI technologies become more entrenched in various societal sectors, it is imperative to delve into the ethical dimensions these technologies entail to ensure they contribute positively to society while minimising potential harm.

Central to this endeavour is the exploration of ethical frameworks for AI, as highlighted by Jobin et al. (2019). This research area is fundamental in crafting and honing guidelines that dictate the use of AI in a manner that benefits society and safeguards against detrimental outcomes. Key considerations within these frameworks include ensuring fairness, upholding accountability, maintaining transparency, and addressing privacy issues that AI technologies may pose. Developing these frameworks is essential in guiding AI technologies towards ethically sound and socially responsible applications.

Impact assessments and developing risk management strategies are other crucial aspects of ethical AI research. This involves thoroughly analysing how AI technologies might affect various domains and proactively identifying strategies to mitigate potential risks. Such assessments are crucial to averting the adverse effects of AI on individuals and communities, ensuring that AI technologies are deployed with foresight and caution.

Furthermore, AI ethics research is instrumental in informing policy and regulation. By providing well-founded insights into the ethical implications of AI, this research can aid policymakers and regulatory bodies in formulating effective and balanced policies that foster innovation while ensuring public safety, privacy, and adherence to ethical norms. This collaborative interface between AI ethics research and policy development is vital for establishing a regulatory environment that supports AI technologies' responsible growth and application.

5.10.2 Engaging the academic community

The engagement between the tech industry and academic institutions is a cornerstone for advancing the field of artificial intelligence, particularly in enhancing AI detection capabilities, exploring ethical dimensions, and comprehensively understanding AI's societal ramifications. Such collaboration is instrumental in pushing the boundaries of current knowledge and technology.

Interdisciplinary research is a fundamental aspect of these collaborations, as underscored by Cath et al. (2017). These collaborative efforts foster a holistic examination of AI's multifaceted impacts by bringing together experts from diverse fields such as computer science, sociology, law, and philosophy. This interdisciplinary approach not only enriches the understanding of AI's societal implications but also contributes to the development of well-rounded solutions to the challenges presented by AI technologies.

Partnering with academic institutions is also pivotal in advancing sophisticated AI detection methods. Through collaborative research, new algorithms and models more adept at identifying AI-generated content, which is becoming increasingly sophisticated and challenging to detect, can be developed. Such advancements are critical in maintaining information integrity and ensuring AI's responsible use in content creation.

Moreover, exploring the broader societal impacts of AI is another vital area where collaboration with academia can yield significant insights. These collaborations can delve into how AI technologies influence various aspects of human life, including employment, education, and social interactions. Understanding these impacts allows for a more informed approach to harnessing AI technologies in ways that benefit society while mitigating potential adverse effects.

Engaging with academic communities through training and workshops is an effective way to disseminate knowledge about AI technologies. These educational initiatives can reach a broad audience, including students, researchers, and industry practitioners, fostering a greater understanding of AI's potential risks, ethical considerations, and best practices. This educational outreach is crucial for cultivating a well-informed community capable of contributing positively to the AI landscape.

5.11 LEGAL AND POLICY MEASURES

5.11.1 Regulatory frameworks

The advent of increasingly sophisticated AI technologies necessitates the establishment of comprehensive regulatory frameworks to oversee the creation and dissemination of AI-generated content. Traditional laws may fall short in addressing the unique challenges presented by these technologies, underlining the need for a more tailored approach.

Crafting specific regulations is a critical step in this process. According to Buiten (2019), it is imperative to devise regulations that directly tackle the intricacies of AI-generated content. This entails defining the boundaries of AI misuse in content creation and distribution and setting explicit guidelines for ethical and responsible AI utilisation. Establishing such regulations ensures clarity around what constitutes acceptable use of AI in content generation, fostering an environment of responsible innovation.

Equally important is the implementation of penalties for the misuse of AI. Establishing clear legal ramifications for creating and distributing detrimental AI-generated content is a significant deterrent against unethical practices.

This component of the regulatory framework is crucial for maintaining the integrity of digital content and safeguarding against the proliferation of harmful materials.

Moreover, regulatory frameworks must be flexible to adapt to the swift pace of technological advancements in the AI domain. This necessitates continuous engagement with AI specialists and technologists to ensure the regulations remain pertinent and practical. Such dynamic regulatory mechanisms are essential for staying aligned with the evolving landscape of AI technologies.

Lastly, protecting individuals' rights and privacy is paramount in formulating AI regulations. The frameworks should be designed to guard against the misuse of AI in activities that could harm individuals, such as creating deepfakes for defamation, harassment, or privacy invasion. By prioritising the protection of personal rights and privacy, the regulatory frameworks can provide a solid foundation for AI technologies' ethical and responsible development and application.

5.11.2 International cooperation

The global nature of digital content and the internet's lack of borders necessitate international collaboration to counteract AI misuse effectively. Such cooperation is crucial for developing a unified strategy to manage the challenges posed by AI-generated content across different regions and legal systems.

One key aspect of this international effort is the standardisation of AI regulations. As Taddeo and Floridi (2018) suggest, international agreements and standards can pave the way for uniform regulations on AI-generated content, ensuring a consistent approach to its misuse. These agreements could cover various aspects, including the ethical use of AI, data protection, and privacy standards, thereby facilitating a cohesive regulatory environment across borders.

Collaborative enforcement efforts are also essential in addressing AI misuse, especially given the cross-border nature of digital content. International cooperation can enhance the effectiveness of enforcement by enabling joint investigations and the sharing of intelligence. Such collaboration is particularly vital for tackling AI-driven online harassment and other forms of digital abuse that transcend national boundaries.

Moreover, international cooperation can provide solutions to jurisdictional challenges that arise in enforcing laws against AI misuse. When the creator of AI-generated content resides in a different jurisdiction from the victim, international collaboration can help bridge the gap, ensuring that legal actions can be pursued effectively across borders.

Finally, engaging in global dialogue and policymaking is instrumental in harmonising approaches to AI governance. Participation in international conferences, treaties, and working groups dedicated to AI can contribute to the establishment of global norms and standards. This collective effort towards policymaking can ensure that strategies for managing AI technologies and their implications are aligned, promoting a more secure and ethical digital environment worldwide.

5.12 TECHNOLOGY USER POLICIES

5.12.1 Robust platform policies

For online platforms to maintain a secure and reliable digital environment, it is essential to have robust policies in place to prevent the misuse of AI-generated content. Gorwa (2019) highlights the importance of establishing clear community guidelines that explicitly address the use of such content. These guidelines should clearly define what constitutes misuse, including the creation of deepfakes for harassment or the dissemination of false information. Providing concrete examples within these guidelines can significantly aid users in understanding the rules and expectations, thereby fostering a more responsible online community.

Equally important is establishing and enforcing consequences for those who breach these policies. Depending on the severity of the violation, repercussions could range from the removal of content and temporary bans to the permanent deactivation of accounts. It is crucial that the application of these policies is consistent and transparent to ensure fairness and maintain trust among users.

To effectively enforce these policies, platforms must implement robust mechanisms. This may involve utilising AI-based detection tools to identify potential policy violations and deploying human moderators to assess and resolve complex cases. Additionally, there might be a need to invest in technology that detects and automatically flags AI-generated content, ensuring a proactive approach to policy enforcement.

Lastly, given the rapid evolution of AI technology, platform policies must not be static. Regular reviews and updates to these policies will ensure they continue to address the latest challenges and forms of misuse associated with AI-generated content. Keeping policies current is vital for effectively mitigating risks and maintaining the integrity of the digital environment. For example, in response to the rising challenge posed by deepfakes and other forms of manipulated media, Facebook has implemented a multifaceted approach to address this issue (Meta, 2020):

1. *Policy Update for Manipulated Media:* Facebook has recently updated its policies to address the issue of deepfakes and other misleading manipulated videos. According to the updated criteria, any videos or content edited or synthesised in ways that are not immediately apparent to an average person and could mislead someone into believing false statements were made will be removed. Additionally, any content

that employs AI or machine learning to alter the content thus making it appear substantially authentic will be taken down.

2. *Exclusions from the Policy:* This policy does not apply to content classified as parody or satire or videos edited solely to omit or change the order of words.
3. *Enforcement of Community Standards:* In line with existing policies, any media, deepfake or otherwise, will be removed if it violates Facebook's Community Standards, which cover nudity, graphic violence, voter suppression, hate speech, etc.
4. *Third-Party Fact-Checking:* Videos that do not meet removal standards are reviewed by independent third-party fact-checkers. If a piece of media is rated false or partly false, its distribution is significantly reduced, and users are warned about its falsity.
5. *Maintaining Public Access for Context:* By not removing all manipulated content, Facebook aims to provide context and awareness about such media, as these videos could still be available on other platforms.
6. *Detection and Removal of Fake Accounts:* Efforts also focus on identifying and removing networks that use AI-generated photos for fake accounts as part of their broader campaign against coordinated inauthentic behaviour.
7. *Deepfake Detection Challenge:* Launched in September, this challenge is a global initiative encouraging research and development of tools to detect deepfakes. Supported by a coalition of organisations, it includes a $10 million grant for participants.
8. *Training for Newsrooms:* In collaboration with Reuters, Facebook has initiated a free online training course to help newsrooms worldwide identify deepfakes and manipulated media, addressing the challenges news organisations face in verifying third-party media sources.

5.12.2 User reporting mechanisms

Empowering users with efficient reporting systems is a crucial strategy for online platforms aiming to foster a secure and respectful environment. Accessible and user-friendly reporting tools are fundamental, allowing users to easily report suspicious or harmful content or violations of platform guidelines. These tools should be seamlessly integrated into the platform's interface, offering straightforward options for users to flag content that might be AI-generated or otherwise breach platform policies.

In addition to providing the necessary tools, educating users on the importance of reporting and correctly using these tools is crucial. This education could take various forms, such as in-app tutorials, comprehensive help articles, or broader educational campaigns, all designed to equip users with the knowledge and confidence to report content effectively.

Platforms must have a responsive review system to promptly assess and act on these reports upon receiving them. This system could blend automated processes for initial screening with human moderation for more complex evaluations, ensuring a balanced and nuanced approach to content review.

Feedback mechanisms for users who report content can further enhance community involvement in maintaining platform integrity. Acknowledging users' contributions through notifications about the status or outcome of their reports or sharing general statistics on the role of user reports in content moderation can motivate continued community engagement.

Moreover, involving the user community in developing and refining reporting mechanisms and policies can significantly enhance the effectiveness of these systems. Platforms might seek direct user feedback, conduct surveys, or organise forums to gather insights into user experiences and perspectives. This collaborative approach fosters a sense of shared responsibility and ensures that reporting mechanisms and policies are aligned with user expectations and needs, ultimately contributing to a safer and more inclusive platform environment.

5.13 ETHICAL AI DEVELOPMENT

5.13.1 Ethical guidelines for AI developers

Ethical guidelines are crucial in guiding the development and application of AI technologies towards responsible use and societal benefit. Hagendorff (2020) emphasizes the significance of accountability in AI development within these guidelines. Developers must be responsible for the outcomes of their AI systems, including unforeseen consequences. This responsibility extends beyond mere compliance with legal requirements to a thoughtful consideration of the broader societal impacts their technologies may have.

Transparency in the functioning of AI algorithms and data utilisation is another cornerstone of ethical AI development. Developers are urged to be forthright about the inner workings of their AI systems and the nature of the data they employ. This should also cover an acknowledgement of the systems' limitations and any potential biases present in the data. Such transparency is essential for fostering trust among users and enabling a well-informed public dialogue regarding AI's role in society.

Moreover, preventing harm is a paramount concern in ethical AI development. Developers are tasked with ensuring that their technologies do not perpetuate or intensify issues such as social inequalities, discrimination, or the spread of misinformation. This necessitates thorough testing and evaluation of AI systems across various real-world settings to detect and address potential adverse effects.

Lastly, developers must consider the broader societal impacts of AI technologies. This includes reflecting on how these technologies might influence employment, privacy, and social interactions. Ethical AI development demands a comprehensive approach beyond technical expertise to include consideration of the social and ethical implications of AI technologies. This holistic perspective is crucial for ensuring that AI serves as a force for good in society, enhancing human welfare and promoting social progress.

5.13.2 Incorporating ethical AI into design

Integrating ethical considerations into the AI design process is essential for fostering responsible usage of these technologies. This involves a multifaceted approach encompassing safeguards against misuse, ensuring transparency, protecting user consent and privacy, and committing to ongoing ethical evaluation.

One fundamental aspect is the creation of AI systems that inherently resist misuse, as Taddeo and Floridi (2018) advocate. This entails designing algorithms to detect and prevent the generation of deceptive or harmful content, such as deepfakes or other forms of synthetic media used for malicious purposes. Embedding such safeguards directly into the AI architecture ensures that the technology is less susceptible to exploitation for unethical ends.

In addition to robust security measures, it is crucial to implement features that facilitate the easy identification of AI-generated content. This can be achieved through digital watermarks or distinct identifiers that signal to users when AI has produced content. Such transparency empowers users to make informed judgements about the authenticity and reliability of the information they encounter.

Another critical ethical pillar is the protection of user consent and privacy. AI systems must be designed to safeguard personal data and comply with stringent privacy standards, such as those outlined in the General Data Protection Regulation (GDPR). This commitment to privacy fosters trust among users and ensures compliance with legal requirements, reinforcing the ethical integrity of AI systems.

Finally, the ethical evaluation of AI technologies should not be viewed as a one-time task but a continuous process. Given the rapid pace of technological advancement and the evolving nature of societal values, it is imperative to regularly reassess AI systems to ensure they adhere to the highest ethical standards. This ongoing evaluation process allows for the adjustment of AI systems in response to new ethical challenges and societal expectations, ensuring their sustained alignment with principles of responsible use.

5.14 REDRESS AND MITIGATION

Addressing the consequences of AI-generated content used for nefarious purposes requires a multi-pronged approach. This approach should focus on supporting victims, holding perpetrators accountable, and implementing preventive measures to avoid future incidents.

5.14.1 Victim support

Support for victims of AI-driven harassment or misinformation is crucial, encompassing both psychological and legal assistance. Matias (2016) emphasizes the importance of providing counselling services to help victims cope with the psychological aftermath of such experiences. These services offer vital support, helping individuals to process their experiences and recover from the trauma. Legal aid plays a complementary role by empowering victims to navigate the complexities of their legal rights and the recourse available, enabling them to take appropriate legal actions against the perpetrators.

Online platforms are responsible for offering remediation services to users impacted by AI-generated content. Gillespie (2019) points out the necessity for platforms to take swift action in removing harmful content, reinstating accounts that may have been unjustly suspended, and providing direct support to those affected. These platforms must have clear guidelines and established response protocols to address such incidents efficiently, ensuring a safe and supportive online environment for all users.

Furthermore, restorative justice approaches offer a valuable method for dealing with the consequences of AI misuse, as Braithwaite (2019) discusses. This approach facilitates a process whereby victims can express the impact of the harm done to them, and perpetrators are allowed to understand the consequences of their actions. This can lead to a mutual understanding and, ideally, the offender's rehabilitation. Restorative justice emphasizes healing and resolution, providing a pathway to address the harm constructively and empathetically.

5.14.2 Accountability of tech companies

Tech companies at the forefront of AI technology development are responsible for ensuring their innovations are not exploited for detrimental purposes. This obligation transcends mere legal compliance, demanding a deep-rooted commitment to ethical practices throughout the entire AI development life cycle.

Integrating ethical considerations into the AI development process from the outset is crucial. According to Hagendorff (2020), this involves a thorough assessment of the potential risks and societal impacts associated with AI applications to mitigate any adverse effects proactively. For instance, in the development of facial recognition technology, companies must weigh the privacy concerns and the potential for misuse in surveillance activities. This proactive approach to ethics in AI development helps foresee and address ethical dilemmas before they escalate into real-world issues.

Moreover, when misuse arises, companies must respond swiftly and effectively. Corrective actions might include revising algorithms to prevent the recurrence of similar issues, eliminating harmful AI-generated content from their platforms, and engaging with those impacted to remediate any damages. An illustrative example of such proactive measures is YouTube's approach to curbing the spread of misinformation on its platform. As stated in the YouTube Official Blog, the platform has updated its algorithms and policies to diminish the visibility of misleading content, showcasing a commitment to counteracting the negative implications of AI misuse (O'Connor & Moxley, 2023).

These steps underscore the critical role tech companies play in fostering the ethical use of AI technologies. By prioritising ethical considerations from the design phase and being prepared to take corrective actions against misuse, these companies can ensure AI serves as a force for good, enhancing societal well-being while minimising potential harm.

5.14.3 Proactive monitoring for misuse

Proactive monitoring plays a vital role in the early detection and mitigation of potential misuse of AI technologies, ensuring these tools are used responsibly and ethically. Investing in effective detection systems is a critical step for tech companies. These systems should be capable of identifying any form of misuse of AI technologies promptly. A notable example is Facebook, which, as highlighted in its Transparency Report (2021), has leveraged AI-based tools to detect and eliminate hate speech and counterfeit accounts from its platform. Such proactive measures help maintain the platform's integrity and create a safer online environment for users.

In addition to sophisticated detection systems, having rapid response mechanisms in place is crucial for addressing any identified misuse efficiently. According to Taddeo and Floridi (2018), this comprehensive response should encompass technical solutions to immediately mitigate the misuse, enforce relevant policies to prevent recurrence, educate users about potential risks and safe practices, and collaborate with legal authorities when necessary. This multifaceted approach ensures that tech companies are capable of detecting misuse and are equipped to respond swiftly and effectively, minimising potential harm and upholding ethical standards in AI usage.

5.14.4 Collaboration with law enforcement

When AI-generated content is utilised for unlawful activities, tech companies must work hand in hand with law enforcement agencies to address these issues effectively. Tech companies can play a pivotal role in aiding investigations by providing essential data and insights. Smith (2020) points out that in situations involving online harassment or cyberbullying with AI-generated deepfakes, these companies can offer invaluable assistance to law enforcement. This can include help in tracing the origins of such content and identifying the individuals responsible for creating and disseminating it. Such collaboration not only facilitates the investigative process but also aids in holding perpetrators accountable for their actions.

However, while engaging in such collaborations, tech companies must balance complying with legal obligations and protecting user privacy. This necessitates ensuring that data sharing and cooperative actions are conducted within the legal framework of privacy laws and are consistent with user consent agreements. Upholding this balance is essential for maintaining user trust and adherence to ethical standards.

Initiatives like the Global Alliance Against Child Sexual Abuse Online (2021) exemplify the effectiveness of joint efforts in combating cybercrimes. This partnership among tech companies and law enforcement agencies underscores the potential impact of collective action against cybercrimes. Such alliances enhance the capabilities to tackle these issues and set a precedent for international cooperation to ensure AI technologies' safe and ethical use.

5.15 POLICY AND LEGAL MEASURES

5.15.1 Legal frameworks for redress

The rapid advancement of AI technologies necessitates a dynamic evolution of legal frameworks to provide victims of AI-generated content misuse with adequate means of redress. AI's unique challenges, such as online harassment, privacy breaches, and digital fraud, demand a comprehensive legal response.

Updating existing laws is a critical step in this process. Many current laws fail to address the intricacies associated with AI, particularly in the context of online harassment, privacy invasion, and digital deception. Buiten (2019) underscores revising such laws to encapsulate the new forms of harm engendered by AI technologies, like those inflicted by deepfakes or AI-generated false narratives. Adjusting defamation laws, for instance, to cover these new harms is essential for providing adequate legal protection.

In instances where existing laws are insufficient, creating new legislation tailored to the specific challenges posed by AI is imperative. This could entail drafting laws that directly target the production and dissemination of AI-generated deepfakes used for malicious purposes such as harassment or impersonation, ensuring that legal systems are equipped to tackle these emerging threats.

Moreover, the global nature of the internet introduces jurisdictional challenges that complicate legal redress for AI-generated content misuse. Effective legal frameworks must, therefore, account for these complexities, potentially requiring international collaboration to ensure that victims have access to justice, even when offences transcend national borders.

A victim-centric approach should be at the heart of legal frameworks dealing with AI misuse. This approach focuses on the needs and rights of victims, ensuring they have access to legal recourse and support. It is crucial that the law not only provides mechanisms for redress but also acknowledges and addresses the specific harms suffered by victims, ensuring their protection and recovery in the face of AI-generated content misuse.

5.15.2 Regulations to prevent repeat offenses

Implementing comprehensive regulations is paramount to mitigate the risks associated with AI misuse and encourage responsible usage of AI technologies. Such regulations not only serve to penalise offenders but also play a preventive role in discouraging potential misuse.

Enacting clear and stringent penalties for using AI technologies can be a powerful deterrent. According to Wachter et al. (2017), legal consequences such as fines, imprisonment, or other punitive measures should be considered for individuals or entities guilty of misusing AI. For instance, regulations could introduce severe penalties for those using AI to create and disseminate harmful, deepfake videos, thus directly addressing one of the more nefarious aspects of AI misuse.

In addition to legal penalties, imposing restrictions or sanctions on repeat offenders can further reinforce the seriousness of AI misuse. Such measures could include prohibitions from using certain AI technologies or platforms and limitations on accessing AI development tools, thereby directly curtailing the ability of offenders to continue misusing AI technologies.

Effective monitoring and enforcement mechanisms are essential to ensure adherence to these regulations. This could involve establishing dedicated governmental bodies overseeing AI usage and compliance or promoting industry-led self-regulation initiatives that encourage ethical practices within the tech community.

Furthermore, regulatory frameworks should emphasise preventive measures alongside punitive ones. This entails mandating AI developers embed safeguards within their technologies to prevent misuse and advocating for ethical AI development practices. By focusing on prevention, regulations can contribute to a culture of responsibility and ethical consideration in the AI field, reducing the likelihood of misuse before it occurs and ensuring a safer and more trustworthy technological landscape.

5.16 PUBLIC AWARENESS AND ADVOCACY

5.16.1 Awareness campaigns

Enhancing public understanding of the challenges associated with AI-generated content and informing individuals about the available support mechanisms for victims are fundamental steps in addressing the misuse of

AI technologies. Educating the public about the potential risks and ethical dilemmas posed by AI-generated content is critical to awareness campaigns. As Wardle and Derakhshan (2017) suggest, such initiatives should elucidate how AI can fabricate deepfakes, manipulate information, or engage in online harassment. To ensure widespread outreach, these campaigns should leverage various channels, including social media platforms, public service announcements, and educational workshops, engaging a diverse audience across different demographics.

Moreover, these campaigns must offer practical guidance on recognising AI-generated content. This involves educating the public on the distinguishing features of manipulated content and emphasising the significance of verifying the sources of information. By equipping individuals with the skills to evaluate content critically, they become more resilient to misinformation and manipulation.

Awareness initiatives should also clearly outline the recourse available to victims of AI misuse. This includes providing comprehensive information on reporting such incidents, the avenues for seeking legal advice, and the availability of psychological support services. Ensuring that victims know these steps is crucial for their recovery and for holding perpetrators accountable.

Collaborations with educational institutions represent a strategic approach to embedding AI literacy within academic curriculums. Such partnerships can foster a more informed and critical future generation adept at navigating the complexities of AI-generated content. Integrating AI education into school and university programmes can lay a strong foundation for digital literacy, preparing students to assess and engage with AI-generated content responsibly and critically.

5.16.2 Advocacy for victims' rights

Advocacy groups play a crucial role in safeguarding the rights of individuals affected by AI misuse, leveraging various strategies to foster a safer digital environment. One of their essential functions is lobbying for policy and legislative changes that specifically cater to the challenges posed by AI technologies. As Citron and Franks (2020) discuss, such efforts can focus on advocating for laws that prevent the unauthorised use of personal data in generating AI content, among other issues unique to AI. Advocacy groups can influence the development of policies and legal frameworks that better protect individuals from the potential harms of AI misuse.

Beyond influencing policy, advocacy groups are instrumental in securing support services for victims. They can champion allocating necessary resources and funding to establish and maintain services such as counselling, legal aid, and online support platforms. These services are vital in providing victims with the assistance they need to navigate the aftermath of AI misuse, from legal recourse to emotional recovery. Advocacy groups can impact the public by raising awareness of AI misuse's ethical implications and damages. Through campaigns, education, and showcasing actual incidents, they can highlight the significance of ethical AI use and the risks of abuse.

Furthermore, advocacy groups can play a pivotal role in promoting the ethical development and use of AI technologies. By collaborating with stakeholders across the tech industry, policymaking, and academia, they can advocate for AI to be developed and utilised in ways that uphold human rights and adhere to ethical standards. This multi-stakeholder approach not only helps set ethical guidelines for AI development but also ensures that these technologies are harnessed for the betterment of society while minimising potential harm.

5.17 CONCLUSION

The intersection of generative AI and online harassment presents a complex challenge, necessitating a holistic and continuously adaptive approach. This strategy must encompass technological innovation, legal frameworks, ethical considerations, and collaborative efforts from various stakeholders. As the landscape evolves, proactive measures and ongoing research are crucial in addressing future challenges and developments.

5.17.1 Developing comprehensive policies and legal frameworks

Crafting nuanced and compelling legal frameworks is fundamental. In collaboration with technologists, legal experts, and advocacy groups, policymakers need to create laws that can evolve with the changing landscape of digital harassment. These laws should specifically address the use of AI in creating and disseminating harmful content, providing clear definitions and guidelines for enforcement. Furthermore, the legal frameworks must be designed to protect victims' rights while not stifling innovation in AI, thereby promoting ethical AI development. Regular reviews and updates of these laws are necessary to keep pace with technological advancements (Buiten, 2019).

5.17.2 Continuous technological vigilance and adaptive detection mechanisms

The advancement of generative AI technologies requires ongoing vigilance in developing countermeasures. This includes enhancing detection algorithms and exploring novel approaches to identify and mitigate AI-driven harassment. Emphasis should be placed on machine learning techniques that adapt to new synthetic media forms and patterns. Collaborative efforts between academia and the tech industry are vital to ensure that these technologies are robust and can respond effectively to new challenges. Additionally, there is a need for research that focuses on the ethical use of detection technologies, ensuring that they do not infringe on privacy or freedom of expression (Zhou et al., 2021).

5.17.3 Global collaboration, standards, and cooperation

The global nature of the internet and AI technologies calls for international collaboration and standard setting. Establishing frameworks that facilitate cooperation across borders, involving governments, tech companies, non-governmental organisations, and academic institutions is imperative. Such collaboration should aim to develop international standards for the ethical development and use of AI, share threat intelligence, and formulate effective responses across different legal and cultural contexts. This global cooperation is essential in developing strategies that are universally effective and respectful of diverse perspectives and laws (Taddeo & Floridi, 2018).

5.17.4 Public awareness, education, and evolving educational programmes

Enhancing public awareness and education is crucial in building resilience against AI-driven harassment. This involves implementing comprehensive educational programmes that cover the risks associated with generative AI and promoting responsible online behaviour. Such initiatives should be diverse, ranging from school curricula to public awareness campaigns and guidelines provided on online platforms. Furthermore, these educational efforts need to be dynamic, evolving with technological advancements to ensure that the public remains informed and capable of critically engaging with AI-driven content. This approach is vital in creating a digitally savvy society that recognises and resists online harassment (Wardle & Derakhshan, 2017).

5.17.5 Ethical considerations in AI development

Integrating ethical considerations into the development and deployment of AI technologies is imperative. This involves prioritising transparency, accountability, and user protection in AI systems. Technology companies and developers must commit to ethical standards that prevent misuse and consider the societal impact of their products. As AI technology advances, it is crucial that ethical frameworks also evolve, addressing new challenges and ensuring that AI developments are aligned with societal values and norms.

This focus on ethics is essential in building public trust and ensuring the responsible use of AI (Hagendorff, 2020).

5.17.6 Proactive approach to emerging threats and evolving tactics

Adapting to the evolving tactics of harassers using generative AI requires a proactive and anticipatory approach. This involves reacting to new threats and actively predicting and preparing for them. Collaborations among cybersecurity experts, AI researchers, and law enforcement are vital to this effort. By understanding the potential future tactics of harassers, strategies and technologies can be developed to counteract these threats effectively. This proactive approach is essential in avoiding harassers and ensuring that countermeasures are effective and timely (Taddeo & Floridi, 2018).

Addressing the challenges of generative AI-driven online harassment demands a comprehensive, multifaceted, and constantly evolving approach. This approach must integrate strong policies, vigilant technology development, global collaboration, public education, ethical AI development, and a proactive stance against emerging threats, ensuring a safer digital environment.

REFERENCES

Birks, D., & Clare, J. (2023). Linking artificial intelligence facilitated academic misconduct to existing prevention frameworks. *International Journal for Educational Integrity*, *19*(1). https://doi.org/10.1007/s40979-023-00142-3

Braithwaite, J., Burford, G., & Braithwaite, V. (2019). Learning from the human services. *Restorative and Responsive Human Services*, 210–234. https://doi.org/10.4324/9780429398704-14

Buiten, M. C. (2019). Towards intelligent regulation of artificial intelligence. *European Journal of Risk Regulation*, *10*(1), 41–59. https://doi.org/10.1017/err.2019.8

Cao, W., Hu, L., & Cao, L. (2015). Deep modeling complex couplings within financial markets. *Proceedings of the AAAI Conference on Artificial Intelligence*, *29*(1). https://ojs.aaai.org/index.php/AAAI/article/view/9559/9418

Cath, C., Wachter, S., Mittelstadt, B., Taddeo, M., & Floridi, L. (2017). Artificial intelligence and the 'good society': The US, EU, and UK approach. *Science and Engineering Ethics*. https://doi.org/10.1007/s11948-017-9901-7

Chesney, R., & Citron, D. K. (2018). Deep fakes: A looming challenge for privacy, democracy, and national security. *SSRN Electronic Journal*. https://doi.org/10.2139/ssrn.3213954

Citron, D. K., & Franks, M. A. (2020). The Internet as a speech machine and other myths confounding section 230 speech reform. *SSRN Electronic Journal*. https://doi.org/10.2139/ssrn.3532691

Darktrace. (2023). *The state of email security in an AI-owered world.* https://darktrace.com/resources/state-of-ai-cyber-security-2024?s_kwcid=AL!14369!3!698960430934!p!!g!!email%20security%20ai&utm_source=google&utm_medium=paidsearch&utm_campaign=AUS-fy24cyberai-bundle-demand&utm_id=combined&gad_source=1&gclid=CjwKCAjw9IayBhBJEiwAVuc3fr-kq2dBNRFFCjsx36tjRKUHFesMafhfv9k0wwtpJnl2ZskDZQeQNRoCXi-IQAvD_BwE&gclsrc=aw.ds#white-paper

De Vries, K. (2022). Let the robot speak! AI-generated speech and freedom of expression. In *YSEC yearbook of socio-economic constitutions 2021*. Springer. https://link.springer.com/chapter/10.1007/16495_2021_38

Dignum, V. (2019). Responsible artificial intelligence. In *Artificial intelligence: Foundations, theory, and algorithms*. Springer International Publishing. https://doi.org/10.1007/978-3-030-30371-6

DiPaola, S., Gabora, L., & McCaig, G. (2018). Informing artificial intelligence generative techniques using cognitive theories of human creativity. *Procedia Computer Science*, *145*, 158–168. https://doi.org/10.1016/j.procs.2018.11.024

Elgesem, D. (2023). The AI act and the risks posed by generative AI models. In *CEUR Workshop Proceedings*. https://ceur-ws.org/Vol-3431/paper3.pdf

Erzberger, A. (2023). WormGPT and FraudGPT – The rise of malicious LLMs. *Trustwave*. www.trustwave.com/en-us/resources/blogs/spiderlabs-blog/wormgpt-and-fraudgpt-the-rise-of-malicious-llms/

European Commission. (2021). Proposal for a regulation laying down harmonised rules on artificial intelligence (artificial intelligence act) COM(2021) 206. https://eur-lex.europa.eu/legal-content/EN/TXT/HTML/?uri=CELEX:52021PC0206

Facebook Transparency Report. (2021). *Facebook*. https://transparency.facebook.com/

Ferrara, E. (2023). *GenAI against humanity: Nefarious applications of generative artificial intelligence and large language models* (arXiv preprint arXiv:2310.00737). https://arxiv.org/abs/2310.00737

Flynn, A., Clough, J., & Cooke, T. (2021). Disrupting and preventing deepfake abuse: Exploring criminal law responses to AI-facilitated abuse. In *The Palgrave handbook of gendered violence and technology* (pp. 583–603). https://doi.org/10.1007/978-3-030-83734-1_29

Gibson, C., & Hagen, J. (2023). How cybercriminals can perform virtual kidnapping scams using AI voice cloning tools and ChatGPT. *Trend Micro*. Retrieved February 29, 2024, from https://www.trendmicro.com/vinfo/au/security/news/cybercrime-and-digital-threats/how-cybercriminals-can-perform-virtual-kidnapping-scams-using-ai-voice-cloning-tools-and-chatgpt

Gillespie, T. (2019). *Custodians of the internet*. Yale University Press. https://doi.org/10.12987/9780300235029

Global Alliance Against Child Sexual Abuse Online. (2021). www.weprotect.org/

Goodfellow, I., Pouget-Abadie, J., Mirza, M., Xu, B., Warde-Farley, D., Ozair, S., Courville, A., & Bengio, Y. (2020). Generative adversarial networks. *Communications of the ACM*, *63*(11), 139–144. https://doi.org/10.1145/3422622

Gorwa, R. (2019). What is platform governance? *Information, Communication & Society*, *22*(6), 854–871. https://doi.org/10.1080/1369118x.2019.1573914

Grossman, M., Grimm, P., Brown, D., & Xu, M. (2023). The GPTJudge: Justice in a generative AI world (Duke Law School Public Law & Legal Theory Series

No. 2023–30). *Duke Law & Technology Review*, *23*(1). https://ssrn.com/abstract=4460184

Guleria, A., Krishan, K., Sharma, V., & Kanchan, T. (2023). ChatGPT: Forensic, legal, and ethical issues. *Medicine, Science and the Law*, *64*(2), 150–156. https://doi.org/10.1177/00258024231191829

Hacker, P., Engel, A., & Mauer, M. (2023). Regulating ChatGPT and other large generative AI models. *Proceedings of the 2023 ACM Conference*. https://dl.acm.org/doi/abs/10.1145/3593013.3594067

Hagendorff, T. (2020). The ethics of AI ethics: An evaluation of guidelines. *Minds and Machines*, *30*(1), 99–120. https://doi.org/10.1007/s11023-020-09517-8

Hao, K. (2021, February 12). Deepfake porn is ruining women's lives. Now the law may finally ban it. *MIT Technology Review*. https://www.technologyreview.com/2021/02/12/1018222/deepfake-revenge-porn-coming-ban/

Henderson, P., Sinha, K., Angelard-Gontier, N., Ke, N. R., Fried, G., Lowe, R., & Pineau, J. (2018). Ethical challenges in data-driven dialogue systems. *Proceedings of the 2018 AAAI/ACM Conference on AI, Ethics, and Society*. https://doi.org/10.1145/3278721.3278777

Jobin, A., Ienca, M., & Vayena, E. (2019). The global landscape of AI ethics guidelines. *Nature Machine Intelligence*, *1*(9), 389–399. https://doi.org/10.1038/s42256-019-0088-2

Kingma, D. P., & Welling, M. (2013). *Auto-encoding variational bayes* (arXiv preprint arXiv:1312.6114). https://api.semanticscholar.org/CorpusID:216078090

Kiritchenko, S., Nejadgholi, I., & Fraser, K. C. (2021). Confronting abusive language online: A survey from the ethical and human rights perspective. *Journal of Artificial Intelligence Research*. www.jair.org/index.php/jair/article/view/12590

Legaltech News. (2024). Tracking generative AI: How evolving AI models are impacting legal. www.law.com/legaltechnews/2024/01/13/tracking-generative-ai-how-evolving-ai-models-are-impacting-legal/

Liu, H.-Y., Maas, M., Danaher, J., Scarcella, L., Lexer, M., & Van Rompaey, L. (2020). Artificial intelligence and legal disruption: A new model for analysis. *Law, Innovation and Technology*, *12*(2), 205–258. https://doi.org/10.1080/17579961.2020.1815402

Livingstone, S., & Third, A. (2017). Children and young people's rights in the digital age: An emerging agenda. *New Media & Society*, *19*(5), 657–670. https://doi.org/10.1177/1461444816686318

Matias, J. N. (2016). Going dark. *Proceedings of the 2016 CHI Conference on Human Factors in Computing Systems*. https://doi.org/10.1145/2858036.2858391

McCarthy, B. (2023, May 22). Fake pentagon explosion image spreads online. *AFP Fact Check*. https://factcheck.afp.com/doc.afp.com.33FV4BU

Meskys, E., Liaudanskas, A., Kalpokiene, J., & Jurcys, P. (2020). Regulating deep fakes: Legal and ethical considerations. *Journal of Intellectual Property Law & Practice*, *15*(1), 24–31. https://doi.org/10.1093/jiplp/jpz167

Meta. (2020, January 7). *Enforcing against manipulated media*. https://about.fb.com/news/2020/01/enforcing-against-manipulated-media/

Miotto, R., Wang, F., Wang, S., Jiang, X., & Dudley, J. T. (2018). Deep learning for healthcare: Review, opportunities and challenges. *Briefings in Bioinformatics*, *19*(6), 1236–1246. https://doi.org/10.1093/bib/bbx044

Mizuta, T. (2016). *A brief review of recent artificial market simulation (agent-based model) studies for financial market regulations and/or rules* (SSRN 2710495). http://mizutatakanobu.com/SSRN-id2710495.pdf

Montasari, R. (2022). Cyber threats and national security: The use and abuse of artificial intelligence. In *Handbook of security science* (pp. 679–700). https://doi.org/10.1007/978-3-319-91875-4_84

National Science and Technology Council. (2016). *The national artificial intelligence research and development strategic plan.* www.nitrd.gov/pubs/national_ai_rd_strategic_plan.pdf

Nguyen, T. T., Nguyen, Q. V. H., Nguyen, D. T., Nguyen, D. T., Huynh-The, T., Nahavandi, S., Nguyen, T. T., Pham, Q.-V., & Nguyen, C. M. (2022). Deep learning for deepfakes creation and detection: A survey. *SSRN Electronic Journal.* https://doi.org/10.2139/ssrn.4030341

O'Connor, J. F., & Moxley, E. (2023, November 14). Our approach to responsible AI innovation. *YouTube Official Blog.* https://blog.youtube/inside-youtube/our-approach-to-responsible-ai-innovation/

Paris, B., & Donovan, J. (2019). Deepfakes and cheap fakes: The manipulation of audio and visual evidence. *Data & Society.* https://datasociety.net/wp-content/uploads/2019/09/DS_Deepfakes_Cheap_FakesFinal-1-1.pdf

Partadiredja, R. A., Serrano, C. E., & Ljubenkov, D. (2020). AI or human: The socio-ethical implications of AI-generated media content. *2020 13th CMI Conference on Cyber security and Privacy (CMI) – Digital Transformation – Potentials and Challenges (51275).* https://doi.org/10.1109/cmi51275.2020.9322673

Poland, C. M. (2023). Generative AI and US intellectual property law. arXiv. https://doi.org/10.48550/arXiv.2311.16023.

Qian, Y., Siau, K. L., & Nah, F. F. (2024). Societal impacts of artificial intelligence: Ethical, legal, and governance issues. *Societal Impacts.* https://doi.org/10.1016/j.socimp.2024.100040

Radford, A., Metz, L., & Chintala, S. (2015). *Unsupervised representation learning with deep convolutional generative adversarial networks.* https://doi.org/10.48550/arXiv.1511.06434

Ramos, S., & Ellul, J. (2024). Blockchain for artificial intelligence (AI): Enhancing compliance with the EU AI act through distributed ledger technology. A cyber security perspective. *International Cybersecurity Law Review, 5*(1), 1–20. https://doi.org/10.1365/s43439-023-00107-9

Rawle, E. (2023, June 15). Deepfakes: There is no such thing as consent. *The One Woman Project.* https://www.onewomanproject.org/consent-1/2023/6/15/deepfakes-there-is-no-such-thing-as-consent

Russell, S., Dewey, D., & Tegmark, M. (2015). Research priorities for robust and beneficial artificial intelligence. *AI Magazine, 36*(4), 105–114. https://doi.org/10.1609/aimag.v36i4.2577

Sall, R. K. (2024, January 2). Another day, another development in artificial intelligence. LACBA News. Los Angeles County Bar Association. https://lacba.org/?pg=lacba-news&blAction=showEntry&blogEntry=100910.

Shroff, L. (2024). AI & copyright: A case study of the music industry. *GRACE: Global Review of AI Community Ethics, 2*(1). https://ojs.stanford.edu/ojs/index.php/grace/article/view/3226

Smith, B., & Browne, C. A. (2020). Tools and weapons: The promise and the peril of the digital age. Penguin Publishing Group. ISBN 9781984877727.
Sorkin, A. R., Warner, B., Kessler, S., de la Merced, M. J., Hirsch, L., & Livni, E. (2023, May 23). An AI generated picture stokes a stock market plunge. *The New York Times*. www.nytimes.com/2023/05/23/business/ai-picture-stock-market.html
Stupp, C. (2019, August 30). Fraudsters used AI to mimic CEO's voice in unusual cybercrime case. *Wall Street Journal*. www.wsj.com/articles/fraudsters-use-ai-to-mimic-ceos-voice-in-unusual-cybercrime-case-11567157402
Taddeo, M., & Floridi, L. (2018). Regulate artificial intelligence to avert cyber arms race. *Nature*, *556*(7701), 296–298. https://doi.org/10.1038/d41586-018-04602-6
Vieira, A., & Sehgal, A. (2017). How banks can better serve their customers through artificial techniques. In *Digital marketplaces unleashed* (pp. 429–440). Springer. https://link.springer.com/chapter/10.1007/978-3-662-49275-8_31
Violino, B. (2023, November 28). AI tools such as ChatGPT are generating a mammoth increase in malicious phishing emails. *CNBC*. www.cnbc.com/2023/11/28/ai-like-chatgpt-is-creating-huge-increase-in-malicious-phishing-email.html
Wachter, S., Mittelstadt, B., & Russell, C. (2017). Counterfactual explanations without opening the Black box: Automated decisions and the GDPR. *SSRN Electronic Journal*. https://doi.org/10.2139/ssrn.3063289
Wakefield Research. (2021). *The new face of identity fraud: How synthetic identities are becoming a massive challenge for businesses*. www.deduce.com/wp-content/uploads/2023/10/Wakefield-Research-White-Paper-for-Deduce.pdf
Wang, P. (2021). *Generative chatbot framework for cyber grooming prevention* (Virginia Tech Electronic Theses and Dissertations). https://vtechworks.lib.vt.edu/handle/10919/107136
Wang, Y., Dantcheva, A., & Bremond, F. (2020). *Deep learning for deepfakes creation and detection: A survey* (arXiv preprint arXiv:2004.11138). https://doi.org/10.48550/arXiv.1909.11573
Wardle, C., & Derakhshan, H. (2017). Information disorder: Toward an interdisciplinary framework for research and policymaking. *Council of Europe Report*, 27, 1–109. https://rm.coe.int/information-disorder-toward-an-interdisciplinary-framework-for-researc/168076277c
Zhou, X., Zafarani, R., Shu, K., & Liu, H. (2021). Fake news: Fundamental theories, detection strategies, and challenges. *ACM Computing Surveys (CSUR)*, *54*(1), 1–40. http://dx.doi.org/10.1145/3395046
Zhu, J. Y., Park, T., Isola, P., & Efros, A. A. (2017). Unpaired image-to-image translation using cycle-consistent adversarial networks. *2017 IEEE International Conference on Computer Vision (ICCV)*. https://doi.org/10.1109/iccv.2017.244

Chapter 6

Trust and risk

Psybersecurity in the AI era

Shasha Yu, Fiona Carroll, and Barry L. Bentley

6.1 INTRODUCTION

Artificial intelligence has penetrated every aspect of our lives in the rapidly evolving landscape of the AI era. From education, healthcare, recruitment, management, shopping, and smart assistants to self-driving cars, artificial intelligence is changing how we work, live, think, and behave.

When we use artificial intelligence to perform tasks for us, it is based on the belief that it is trustworthy and that the risks, if any, should be manageable. In other words, the benefits outweigh the risks. In the past, some scholars categorised security threats brought by AI into digital security, physical security, and political security (Bundage et al., 2018). Other researchers have expanded the scope of security threats to include infrastructure, endpoint, network security, security operations and incident response, threat intelligence, mobile security, and human security (Jain, 2021). These studies typically focus on protecting physical assets, digital data, or systems. It involves strategies and measures that can be objectively assessed, such as technological tools, surveillance, and physical barriers.

Some recent research has focused on individuals' perceptions, experiences, and well-being in AI environments (Chattopadhyay & Beyene, 2023; Franklin & Chattopadhyay, 2021). This has given rise to an emerging topic and research area called 'Psybersecurity' (Chattopadhyay & Beyene, 2023). Psybersecurity goes beyond the traditional security paradigms, aiming to safeguard mental health by addressing the psychiatric implications of technology use. It involves an examination of the mental health attack surface and vectors within the broader context of human security (Franklin & Chattopadhyay, 2021).

6.1.1 The emerging importance of securing mental health in the digital age

The relationship between mental health and the use of technology is a complex and multifaceted topic. Mobile device ownership and social media usage have reached unprecedented levels, and concerns have been raised that this constant connectivity harms people's mental health (Odgers & Jensen, 2020). Moreover, people who have experienced cybersecurity attacks frequently

DOI: 10.1201/9781032664859-6

endure negative mental health impacts, such as anxiety and depression, to name just two. In fact, cyberattacks can have devastating consequences for vulnerable people, those close to them, and many other stakeholders (Ives, 2016). However, it seems to be a bit of a chicken-and-egg scenario.

On the one hand, technology can contribute to the development of mental health issues, whilst on the other, people are also turning to technology as a means to address and alleviate these very challenges. For example, accompanying the rising rates of reported mental distress during the COVID-19 pandemic, there was also a reported increase in the use of digital technologies to manage health generally and mental health more specifically (Abd-Alrazaq et al., 2021). Considering this, and particularly as AI technologies continue to advance, taking proactive steps to prioritise mental health in the digital age is essential for the well-being of society.

6.2 TRUST AND RISK IN AI

Understanding the psychological foundations of trust is crucial for the sustainable development of AI technologies. Building and maintaining trust requires a combination of elements and may vary depending on the specific circumstances of the technology and people involved. As researchers (Gulati et al., 2019) point out, the more the tasks get delegated to intelligent systems, the more the nature of user interactions with them becomes increasingly complex, and the more critical it is to understand the amount of trust that a user is willing to place on such systems. Indeed, user trust in AI-enabled systems has been increasingly recognised and proven as a critical element in fostering adoption (Bach et al., 2022). Despite its great theoretical importance, a limited amount of research has still examined how and why trust develops, is maintained, and occasionally unravels in relationships (Simpson, 2007). From the AI perspective, there is also a risk of bias and lack of fairness, misinterpretation of data, lack of transparency and over-reliance on AI, and security vulnerabilities, to name but a few. Balancing trust and managing risks in AI requires a combination of technical and ethical as well as a commitment to continuous improvement and accountability. Nowadays, organisations must actively work towards building and maintaining trust while mitigating potential risks.

6.2.1 Psychological foundations of trust

Trust is a complex, multidimensional construct, making it difficult to operationalise, measure, and interpret (Simpson, 2007). Indeed, there is more than one way to define and describe trust. As Simpson (2007) shows, 'trust can be construed in different ways, and it might have varied importance at different stages of relationship development'. Moreover, user trust in AI-enabled systems is influenced by three main themes: socio-ethical considerations, technical and design features, and user characteristics (Bach et al., 2022). The theoretical concept of trust is fundamental to the successful design of intelligent technologies (Raats et al., 2020).

Researchers, such as Kaewkitipong et al. (2022), have even found that trust plays a more significant role than human–computer interaction (HCI). The psychological foundations of trust can refer to cognitive and emotional factors that lead to a person's willingness to have confidence in a person and/or entity (e.g., technology). This trust is often built when a person and/or technology consistently behaves reliably. As Zenger and Folkman (2019) note, consistency is an essential element of trust for leaders (e.g., they do what they say they will do, etc.). It is true to say that when a person's or a machine's actions and behaviours align with people's expectations, trust tends to grow. Moreover, trust can be influenced by the person or technology's perceived competence, knowledge, and expertise. If a technology consistently affords competence, it is more likely to be trusted.

Regarding leadership, a decisive factor in whether people trust a person as a leader is the extent to which a leader is well-informed and knowledgeable (Zenger & Folkman, 2019). Trust is also closely tied to perceptions of honesty and integrity. For instance, Gefen et al. (2003) found that online trust is built through (1) a belief that the vendor has nothing to gain by cheating, (2) a belief that there are safety mechanisms built into the website, and (3) by having a typical interface, (4) one that is, moreover, easy to use. Trust is more likely to develop when individuals believe that applications are truthful and transparent and act with integrity.

Moreover, trust can be enhanced when people feel understood and supported. Demonstrating empathy and showing that the application is designed to care about the well-being of people can foster trust. Regarding humans (leaders), trust is partly based on how much a leader can create positive relationships with other people and groups (Zenger & Folkman, 2019). Emotions also play a crucial role in how people trust technology and other people. As Ajenaghughrure et al. (2018) say, trust and emotion are critical indicators for persuading users to change their behaviour towards a technical artifact. Establishing an emotional connection, such as through shared values or experiences, can deepen trust. Previous experiences with a person or technology can significantly shape trust. For example, in online banking, the service value affects customer loyalty to the bank through trust and the bank's reputation (Burhanudin, 2022). Positive experiences can build trust, while negative experiences can destroy it.

Finally, open and transparent communication is also fundamental for trust. As Wiencierz and Lunich (2020) show, the more the information users obtain regarding open data, the more trustworthy they perceive the app provider to be and the more likely they are to use the app. Trust is more likely to flourish when information is freely shared, and there are no hidden agendas. Trust is also nurtured when people consistently meet commitments and fulfil their obligations. In their study, Alhaji et al. (2021) detected four factors while trust accumulates (perceived reliability, perceived dependability, perceived predictability, and faith) which do not appear while it dissipates. Furthermore, dependability can be described as a foundation for trust.

6.2.2 Context and significance of trust and risk in the AI era

Trust plays a vital role in the psychological factors affecting security in AI. Trust forms the cornerstone of successful AI adoption, impacting users' willingness to engage with AI technologies and shaping their expectations of reliability, transparency, and ethical behaviour. In April 2019, the Ethics Guidelines for Trustworthy AI were adopted by the European Commission's High-Level Expert Group on AI, emphasising that trust in the technology is essential for people to fully and confidently harness the advantages of AI (European Commission & Content Directorate-General for Communications Networks, and Technology, 2019).

However, trust is a multifaceted concept, encapsulating the belief that another party will act with benevolence, integrity, competence, and predictability (McKnight & Chervany, 2001). When people trust another party, it implies they are willing to take risks, such as being vulnerable, in a relationship (Schoorman et al., 2007).

Nevertheless, the complexities and potential limitations of AI systems introduce various risks. These risks include bias and fairness, data privacy breaches, lack of interpretability, unintended consequences, and the potential for malicious use of AI. If, for any reason, the AI product or device fails to act with benevolence, integrity, competence, and predictability as expected, it may pose risks to individuals and adversely affect mental health.

6.2.3 Benefits of trust in AI

Trust in AI significantly improves efficiency at the individual level by expediting decision-making, reducing cognitive load, automating routine tasks, enhancing time management, optimising resource allocation, facilitating learning, enabling proactive issue resolution, and boosting overall confidence and job satisfaction. Decisions made automatically by AI were often evaluated as being on par with, or even superior to, those made by human experts in certain circumstances (Araujo et al., 2020). Trust in AI systems leads to greater confidence in their decisions and recommendations. Individuals can make decisions more quickly and with less hesitation, resulting in streamlined workflows and faster progress. A study demonstrates that ChatGPT, a widely used large language model–based chatbot, significantly improves average productivity: the time taken decreases by 0.8 standard deviations (SDs) and the output quality increases by 0.4 SDs (Noy & Zhang, 2023). Previous studies show that employees' trust in AI helps motivate collaboration with AI, thus benefiting career well-being and productivity (Kong et al., 2023).

6.3 RISKS STEMMING FROM TRUST IN AI

While vital for its successful integration and widespread adoption, trust in AI carries inherent risks that can have significant societal and economic implications. Trust will impact users' willingness to adopt AI products. For instance, individuals are more inclined to embrace driverless vehicles after receiving positive information about them but less willing to do so after hearing negative information about them (Anania et al., 2018). This can lead to a mismatch between the user's trust in the AI and the AI's trustworthiness. These risks manifest as over-trust and under-trust in AI products, leading to distinct challenges. Disuse, misuse, and abuse are dangerous when trust and trustworthiness are misaligned. Specifically, trust exceeding trustworthiness leads to misuse, while trustworthiness exceeding trust leads to disuse (Parasuraman & Riely, 1997). Figure 6.1 illustrates the interplay between a user's trust in AI and AI's trustworthiness, influencing the user's engagement with AI.

Optimal utilisation and benefits from AI products and services occur when users' trust aligns with the AI's trustworthiness, resulting in a healthy use of AI (Figure 6.1). However, if an AI system exhibits low trustworthiness due to inherent unreliability or malicious third-party use, excessive trust from users can lead to misuse or potential abuse (Figure 6.1).

Conversely, when users lack trust or have lower trust than the AI system's actual trustworthiness, this under trust may result in the underutilisation or disuse of AI (Figure 6.1). The misalignment between user trust and AI trustworthiness introduces risks in over-trust and under-trust scenarios.

In the context of AI environments, we can define misuse as the excessive dependence on automation, such as using AI when it is not appropriate or failing to supervise its functions effectively. Disuse would be the inadequate utilisation of automation, such as disregarding or deactivating AI-driven alerts or safety mechanisms. Abuse refers to the inappropriate design or management of AI systems, where AI is applied without due consideration for its impact on human performance within the resulting system (Parasuraman & Riely, 1997).

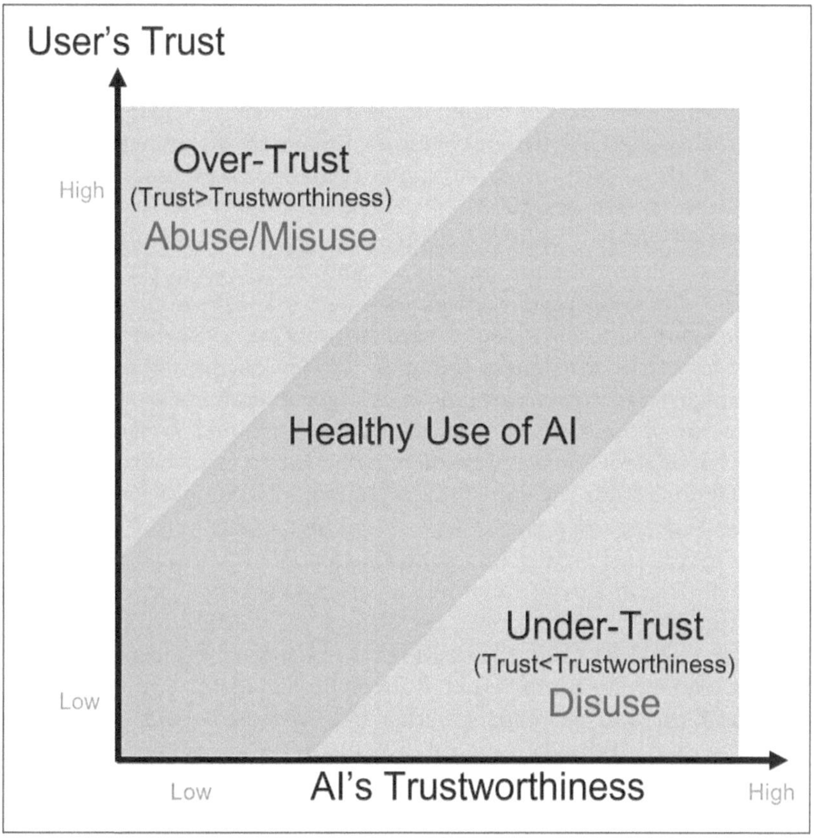

Figure 6.1 Risks Stemming from Trust in AI.

6.3.1 Over-trust in AI products: abuse and misuse

While trust in AI offers numerous benefits, it can also introduce risks if there is an over-trust in AI applications or devices. One significant risk is the potential misuse of AI. Over-trusts in AI can lead to emotional dependence and privacy risk. Positive experiences with AI applications or devices can elicit feelings of affection, including perceived intimacy, passion, and commitment (Hernandez-Ortega & Ferreira, 2021). Experiences characterised by perceived intimacy, enjoyability, and commitment can give rise to para-friendships (Chung-Wa et al., 2020). The introduction of anthropomorphism, in particular, poses significant risks, as individuals may place undue trust in a robot (Lewis et al., 2018).

Interestingly, a study showed that when participants were asked to choose the correct answer for complex mathematical problems provided by different robots, they tended to trust a robot with specific facial aesthetics, even though it provided the wrong answer (Pinney et al., 2022). These emotions might cause individuals to overlook the privacy risks of using and communicating with AI. For example, some people may expose their private information to ChatGPT without being aware that their chat might be stored for up to 30 days, used for model training, and, in some cases, exposed to human reviewers (Hunter, 2023).

Over-trust in AI can reduce vigilance and neglect of potential risks and threats. AIs inherently have limitations and deficiencies, and blind trust in them may cause people to overlook these risks. Previous studies have shown that individuals with solid trust in others tend to dismiss evidence of potential untrustworthiness (Holmes, 1991). Placing blind trust in AI systems might lead to overlooking security vulnerabilities, making them susceptible to cyberattacks and breaches. For example, Samsung workers utilised ChatGPT for code generation and debugging, inadvertently integrating confidential company information, potentially compromising privacy, and illustrating the risks associated with undue reliance on AI technologies (Maddison, 2023). Researchers found that people often over-trust robots, with individuals and groups granting a robot access to a locked dormitory door about 70–80% of the time, even when they suspect it might pose a bomb threat (Booth et al., 2017). Researchers studied human reactions to a robot's guidance in an experiment involving simulated fire emergencies. Surprisingly, 95% of participants either followed the robot's directions or remained with it, even when the robot made mistakes, as they believed the robot was more knowledgeable or incapable of leading them astray (Robinette et al., 2017). In particular, the more accurate the AI system, the more likely it is to be trusted, to the extent that even professionals such as police officers may abandon professional intuition in favour of machine results (Kearns & Muir, 2019).

In addition, over-trust in AI can lead to dependency and a lack of critical thinking. Many people trust AI to make more informed decisions than they can. A study of 1,506 participants reveals that co-writing with opinionated language models affects users' views (Jakesch et al., 2023). In this context, if the AI-generated information is inaccurate or false, it can lead to harmful outcomes for users. Researchers have investigated that ChatGPT can generate fake peer-reviewed citations and references that do not exist in the journal to which they are attributed or anywhere else (Day, 2023). AI-generated information looks so trustworthy that even highly trained professionals and experts can be misled by it. For example, two lawyers were sanctioned for using fictitious case citations generated by ChatGPT in a legal brief (Merken, 2023).

Furthermore, over-trust can prevent a critical examination of AI algorithms, inadvertently perpetuating biases and discriminatory practices. A researcher tested a ChatGPT detector from Turnitin, a software already utilised by 2.1 million teachers to identify plagiarism. It falsely flagged an innocent student, potentially causing trouble (Fowler, 2023). Furthermore, the AI text detectors displayed bias against non-native English speakers, inadvertently flagging work produced by those who use a limited range of words and phrases (Williams, 2023).

Another notable risk is the potential abuse of AI. When individuals interact with others through customisable AI assistants, they may configure their agents to engage in less ethical behaviour and increase deception compared to their actions (Mell et al., 2020). Research has shown that people are more willing to deceive through AI, receiving less blame for their AI's negative behaviour and predicting less blame when using AI for their actions (Gratch & Fast, 2022). On the other hand, AI technology allows people without expertise to easily create professional images, audios, videos, and texts that only professionals could create in the past. This could lead to the abuse of AI, causing harm to others, such as the creation and dissemination of fake content, furthering misinformation and causing damage to individuals and society. As an illustration, DeepTrace, an artificial intelligence company, identified 15,000 Deepfake videos online in September 2019, marking a nearly twofold increase within nine months (Sample, 2020). Significantly, image-based abuse, involving the creation or modification of an individual's image without their consent, disproportionately impacts women. According to a Deepfake research report, it was revealed that approximately 96% of all online Deepfake videos were categorised as Deepfake pornographic videos (Patrini, 2019). These Deepfake images or videos use photos from the internet or social media without the victim's consent or even knowledge, not only hurting the victim emotionally and psychologically but also damaging their reputation. The victims impacted by these Deepfakes may feel horrified, mortified, devastated, ashamed, humiliated, embarrassed, frightened, and paranoid (Melville, 2019).

6.3.2 Under-trust in AI products: disuse

Although AI's prevalence and various applications have increased substantially, many individuals still struggle to trust this technology (Gillath et al., 2021). A study showed that only half of the respondents said they trust companies that use AI as much as they trust other companies (Myers, 2022). Insufficient trust in AI products can result in disuse, hampering potential benefits and opportunities (Lee, 2008). Another study revealed that only one-third of respondents rely on AI for various entertainment-related recommendations or would let it produce goods made by automated machinery. Even more concerning, only 9.3% of consumers would allow AI to manage their finances, and merely 4% would trust it with HR-related work. Additionally, the percentage of those who trust medical diagnostics made by learning and decision-making algorithms is even smaller (Dujmovic, 2017).

People may adapt their use of AI based on changing levels of trust, potentially adding or removing tasks from AI workflows or adjusting their reliance on the AI (Dorton et al., 2022). With the prevalence of large language models (LLM), AI now causes more concern for its potential threat to humans. A group of influential computer scientists and tech industry figures has raised alarm about the potential dangers of highly advanced AI systems and called for a six-month pause in training AI models more powerful than GPT-4, citing concerns for society and humanity (O'Brien, 2023). Following this call, additional concerns about the threats posed by AI have been voiced, with some arguing that a longer pause is needed (Noonan, 2023) and others going further to suggest shutting down large AI training development worldwide, advocating for an indefinite moratorium (Yudkowsky, 2023).

However, it is worth noting that the disuse of AI due to under-trust can hamper potential benefits and opportunities, such as inefficient operations, missed opportunities, reduced competitiveness, and customer dissatisfaction. Understanding and mitigating these risks associated with trust in AI are crucial for establishing a balanced and informed approach to AI integration. Striking the right balance between trust and vigilance is essential to harness AI's benefits while mitigating potential harms and ensuring its responsible and beneficial use in society.

6.4 PSYBERSECURITY RISK FRAMEWORK

As previously discussed, the misalignment between people's trust in AI and the actual trustworthiness of AI can significantly influence attitudes and behaviours towards AI. This misalignment may result in disuse, misuse, or abuse of AI, leading to associated risks. Figure 6.2 illustrates these risks.

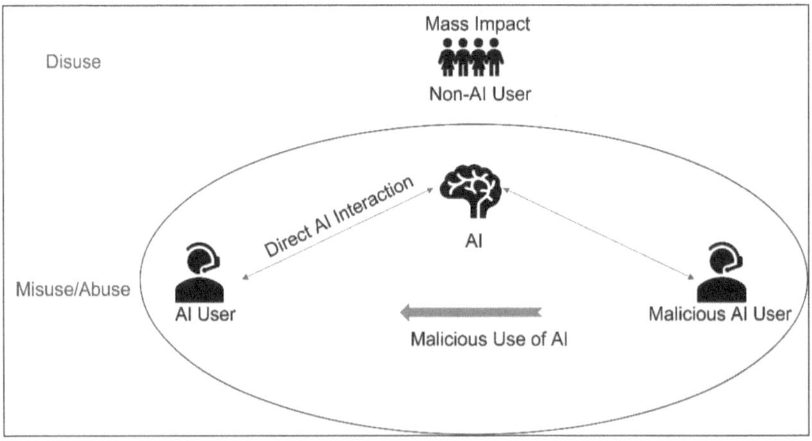

Figure 6.2 Psybersecurity Risk Framework.

In the era of AI, its impact is pervasive, affecting everyone, whether or not they are directly using AI. It is crucial to emphasise the risks of mass impact when discussing Psybersecurity in the AI era; these risks are situated within the 'Disuse' layer. Simultaneously, psychological impacts stem from users' direct interactions with AI systems or third parties' malicious use of AI. These factors constitute the risks within the 'Misuse/Abuse' layer.

6.4.1 Mass impact

In our modern society, where AI algorithms play a prevalent role in various aspects of our lives, everyone – even those who may not be directly using AI systems – is potentially impacted by AI to varying degrees.

While people enjoy the benefits brought by AI, such as increased efficiency and productivity, they also face the threats and tensions introduced by AI, including the risk of job displacement (World Economic Forum, 2020). According to a PwC study, the progression of AI towards greater autonomy could lead to potential automation risks, affecting approximately 30% of jobs and 44% of workers with lower educational backgrounds by the mid-2030s (Hawksworth et al., 2018). Similarly, the World Economic Forum anticipates that the impact of artificial intelligence and automation may replace 85 million jobs by 2025. However, they also project the creation of 97 million new jobs across 26 countries during the same period (World Economic Forum, 2020). This impact is not limited to non-skilled jobs; even highly professional work faces challenges.

An illustrative example is the recognition of AI-generated artwork winning the top prize in an art competition, showcasing AI's popularity at a lower cost and quicker response times than human artists (Roose, 2022). This may have a potential psychological impact on individuals with concerns about job displacement by AI, economic uncertainties, and fear of changes in the workforce. Furthermore, many AI-driven applications are developed by extracting data from the open web, teaching algorithms to recognise patterns and relationships, and generating new work in the same style. Professionals who upload their work to the internet may inadvertently contribute to training their algorithmic competitors. This can lead to feelings of anxiety, insecurity, and a loss of pride in one's work, causing demoralisation and a sense of meaninglessness among specific segments of the population.

Nowadays, AI algorithms are increasingly being utilised for decision-making. Algorithmic prejudice is a factor that can lead to unfairness and have adverse effects on mental health. Psychological harm may result from biased algorithms, leading to unfair or discriminatory outcomes for individuals affected by algorithmic decisions.

AI learns from the data it is fed, making the data quality critical to its effectiveness. Biased or contaminated data can lead to biased outputs from artificial intelligence. For instance, when AI recommends jobs to jobseekers based on their information, it might suggest lower-paying jobs to women due to gender wage gap biases in the database, even if the jobseeker has an excellent educational and professional background (Dastin, 2022). Such biases can make individuals feel ignored and humiliated, leading to psychological harm, including feelings of anger, helplessness, self-doubt, and lowered self-esteem. Algorithms built upon biased datasets or designed with inherent biases can perpetuate and exacerbate existing societal prejudices.

Furthermore, algorithmic bias can contribute to social isolation and stigmatisation. For instance, during the COVID-19 pandemic, China implemented health codes to label individuals based on risk factors, assigning a red code to those needing immediate quarantine (Yu & Carroll, 2023b). If certain groups consistently face unfair treatment or discrimination due to algorithmic decisions, they may experience feelings of frustration, helplessness, marginalisation, diminished self-worth, or exclusion from societal opportunities.

In recent years, AI technologies have been increasingly employed to generate and disseminate news and information, contributing to the proliferation of fake news and misleading information. An investigation into verified true and false news stories distributed on Twitter reveals that falsehoods spread significantly farther, faster, more profound, and more broadly than the truth across all categories of information (Vosoughi et al., 2018). Although more novel than actual news, false news inspires fear, disgust, and surprise in replies. In contrast, true stories evoke feelings of anticipation, sadness, joy, and trust (Vosoughi et al., 2018). Additionally, exposure to false or misleading information generated by AI may lead to psychological distress, such as distorted perceptions of reality.

AI algorithms rely on training with vast amounts of data, much of which is collected from the open web. This raises concerns about the potential breach of users' personal data without their awareness. Large AI training datasets, such as the LAION-5B dataset with over 5 billion images, raise privacy concerns (Xiang, 2022). These datasets gather diverse internet data without individuals' awareness, including images of living artists, photographers, medical content, and people who may not expect their data to be used for AI training. Notably, individuals cannot opt out of inclusion in these datasets.

Additionally, some datasets contain disturbing content like manipulated celebrity porn, non-consensual porn, and graphic images, including those of ISIS beheadings (Xiang, 2022). In addition, AI excels at inferential analytics, which involves generating derived or inferred data from a combination of various pieces of information (Wachter & Mittelstadt, 2019). This capability has the potential to pose a threat to people's privacy as more data is collected and aggregated.

Psychological threats may arise from the widespread use of mass facial scanning and surveillance networks in public places. Various biometric information, such as voiceprints, gait, fingerprints, and faces, can be captured without individuals being aware, leading to breaches of personal privacy. Facial recognition systems can accurately identify people even when they wear masks (Singh, 2017). Additionally, high-resolution cameras can identify faces and license plates in photographs taken from several kilometres away (Schneier, 2019). The surveillance of people's behaviour can increase psychological considers, manifesting as anxiety and distrust among citizens about how their data is used. This sense of being under suspicion or feeling unsafe may lead to self-censorship, creating a chilling effect. Surveillance is also associated with behavioural changes, including fear of expression, avoidance of public gatherings, and reluctance to exercise the right to protest due to fear of reprisals (Starr et al., 2008).

6.4.2 Direct AI impact

Compared to non-AI users, individuals who actively engage with AI technologies may experience more significant psychological impacts from AI. While moderate use of AI is beneficial to people's lives and work experiences, excessive reliance on AI can lead to psychological harm, such as dependency issues, reduced critical thinking, and potential addiction.

With the increasing prevalence of AI tools in education, they play a dual role. On the one hand, these tools assist individuals in mastering various skills and expanding their abilities. On the other hand, they introduce challenges and potential risks (Becker et al., 2023). AI tools can potentially make students overly dependent on them, fostering overconfidence that may impede individual learning and training. This over-reliance can lead to students facing difficulties performing as expected once they no longer use these tools, consequently hindering the development of self-confidence (Yu & Carroll, 2023a).

AI in social media, while facilitating efficient communication and providing users with personalised content, can have psychological ramifications for users. Many major tech companies utilise psychographic analysis for customer profiling. Psychographic analysis seeks to examine the subjective attributes of users to comprehend or forecast their actions. This involves scrutinising personality traits, activities, interests, opinions, needs, goals, values, and attitudes (Wells, 1975). Typically, this information is acquired through the analysis of diverse behavioural data. For instance, researchers can unveil users' opinions, emotions, and attitudes by examining their social interactions on social networks and their sentiments (Chen et al., 2022). AI-driven psychographic analysis can accurately predict a user's preferences and potential future actions based on their past behaviour or the conduct of other users within the same cohort.

Additionally, it can deploy tailored engagements, such as displaying ads, delivering news, or suggesting connections with people they are likely to connect with. Based on psychographic analysis, the personalised content promoted by social media platforms has the potential to create echo chambers, reinforcing users' existing beliefs and limiting exposure to diverse viewpoints (Cinelli et al., 2021). This becomes especially concerning when negative thoughts are exposed on social media, as they may be further entrenched and reinforced. Previous studies have indicated a correlation between social media use and mental health issues, including depressive disorders and suicidal behaviour, particularly among young people (Sedgwick et al., 2019). Additionally, research has demonstrated a link between comments on Instagram and the escalating severity of self-injury, suggesting that social media platforms may inadvertently contribute to the reinforcement of harmful behaviours (Brown et al., 2017). In addition, algorithmic distortions can lead to distorted perceptions of reality and self-worth, resulting in negative consequences such as the spread of misinformation or manipulated behaviour. For instance, young people may be susceptible to suicidal behaviour when exposed to information about the suicidal behaviour of others (Dunlop et al., 2011; Haw et al., 2013).

In addition, concerns about supernormal stimuli arise in AI and technology, suggesting that individuals may become overly attracted to and immersed in artificial stimuli designed to be more enticing or rewarding than real-life experiences (Luxton & Watson, 2023). This concern is particularly pronounced with AI technologies such as virtual reality environments and chatbots, which can generate rich and engaging content. Overexposure to these extraneous stimuli raises concerns about the social and emotional well-being of individuals, potentially leading to reduced connectedness in real-life relationships and adverse mental health outcomes. This may include feelings of isolation and disconnection from the tangible world. It can also distort the user's perspective on interpersonal relationships. For example, instructing a digital assistant to perform tasks on one's behalf can create the illusion of having power over another person (Fast & Schroeder, 2020).

The COVID-19 pandemic has profoundly transformed the way people work and live globally. We have witnessed a growing shift in learning, shopping, and socialising from offline to online (Caroppo et al., 2021). Technologies associated with the metaverse, such as augmented reality (AR), virtual reality (VR), and mixed reality (MR), are introducing real-world experiences into new virtual spaces. Virtual reality technology utilises various hardware and software tools that interact with our senses (sight, hearing, touch, smell, sense of balance, etc.) to construct an environment recognised as real by people's bodies and minds (Dremliuga et al., 2020). However, the experience of being in a virtual environment can have tangible effects on the body and mind. Studies have indicated that individuals in VR may exhibit fear when walking across a plank of wood on the ground in the real world, even if it is placed between virtual roofs (Wang et al., 2019). Virtual reality users may even experience physical harm, such as a heart attack induced by a game (Cleworth et al., 2016). In addition, individuals may become overly immersed in AI-created virtual environments, potentially leading to detachment from reality.

6.4.3 Malicious use of AI

With the increasing intelligence of AI, AI-driven applications have become widely accessible, exposing people to the malicious use of AI and AI-enabled crimes (Blauth et al., 2022). AI algorithms heavily depend on data training, creating vulnerabilities that can be exploited for nefarious purposes. Precisely, injecting specially crafted adversarial data into a target training dataset can manipulate machine learning results – a technique known as data poisoning (Wang & Chaudhuri, 2018). An illustrative example involves perturbed data altering the results of self-driving cars' recognition of traffic signs (Ding et al., 2019). Detecting data poisoning is challenging and may leave a backdoor for manipulation if AI models adopt the poisoned data (Chen et al., 2017). Consequently, individuals affected by AI trained on poisoned data may be misled or manipulated, leading to unpleasant or unfair experiences that could be potentially harmful to their well-being.

In addition, AI has been increasingly used for fraud or deception. One example is the misuse of deepfakes for deception, where artificial intelligence is employed to create and disseminate convincingly fake or misleading images, videos, and media. This technology poses serious risks by manipulating content in ways that are challenging to detect, potentially leading to deception or misinformation. Instances have been reported where malicious individuals use Deepfake-generated composite portraits and fake social media profiles to approach and connect with thousands of people on LinkedIn. When a target accepts these connections, other users may perceive it as an endorsement, potentially lowering their guard (Raphael, 2021). The negative impacts of AI-enabled deception can lead to feelings of betrayal, mistrust, and fear, with detrimental effects on mental health. Some criminals exploit AI software for illicit purposes, such as using voice synthesis software to mimic the voices of victims' friends and relatives for fraudulent activities. This results in financial losses, undermines interpersonal trust, and instils a sense of insecurity, causing emotional distress. Concerns arise about using deepfakes to spread misinformation or propaganda, contributing to a decline in trust in information and institutions. This erosion of trust can lead to breaking social norms and values and increased social polarisation.

AI can also be exploited for social engineering attacks. A study shows that social robots could employ psychological manipulation techniques to deceive people into providing private information without their awareness (Abate et al., 2020). Chatbots can also be manipulated for malicious purposes. For instance, employing a method called DAN, or 'Do Anything Now', researchers have successfully bypassed ChatGPT's safety features, triggering the generation of content that violates OpenAI's policies against violence, offensive material, and sexually explicit content. There have also been reported cases of Bing being exploited to threaten users (Perrigo, 2023).

6.5 DISCUSSION

Based on the preceding discussion, addressing the risks and psychological impact arising from the misalignment of people's trust and AI's trustworthiness can be approached through a user-centric approach and an AI-centric approach. The user-centric approach focuses on increasing users' awareness of AI's risks and threats, cultivating a balanced trust in AI, and ultimately using AI in moderation. Promoting a better understanding of AI and cultivating appropriate risk awareness through educational initiatives and awareness training is crucial. Comprehensive awareness training enables users to assess the potential risks and impacts associated with AI accurately. This, in turn, promotes the healthy and informed use of AI and mitigates the potential for disuse, misuse, or abuse resulting from over- or under-trust due to misconceptions.

The AI-centric approach focuses on increasing trustworthiness from technical, ethical, and legal perspectives. From a technical perspective, increasing transparency can help reduce bias or errors and increase the reliability and trustworthiness of AI. From an ethical perspective, creating and refining an industry standard could increase accountability and benefit the created responsible and ethical AI. Moreover, a robust legal framework is the bottom line for avoiding and defeating malicious or illegal creation and use of AI.

6.5.1 Awareness training

As discussed above, proper awareness is crucial to using AI to benefit people's lives in modern society. Both over-trust and under-trust can hinder the healthy use of AI. Awareness training is essential in today's AI era, not only for AI professionals and AI users but also for every ordinary person, because AI is indeed affecting everyone's lives, even if they do not directly use AI. The public should have the right to know if their data is being used for AI training or decision-making. They should also be informed if an AI is being used for a purpose related to them and what steps they can take to mitigate the impact.

Perceptions of risks, whether accurate or not, significantly impact trust in AI (Sanders et al., 2011). Users should understand the potential vulnerabilities, security threats, and unintended consequences of AI and the ethical and psychological implications of AI technologies. People can trust and use its capabilities better when they understand how AI works.

Education on the proper use of AI is critical for the public, especially youth, to promote a balanced use of AI technologies. This education aims to prevent over-dependence and mitigate potential adverse psychological effects associated with dependency. Managing the psychological impact of immersive virtual worlds by promoting healthy boundaries and awareness of virtual experiences is essential for informed decision-making and effective use of AI tools.

6.5.2 Transparency in AI

Transparency in AI is crucial in artificial intelligence, serving as the bedrock for building trust and mitigating risks associated with AI systems. It is pivotal in enhancing trust among AI developers, users, and the broader public. When individuals comprehend how AI systems operate, make decisions, and draw conclusions, they are more inclined to place trust in the technology.

AI autonomously learns from existing data, identifying patterns and rules for new data. While users are familiar with the inputs and outputs, the internal process remains opaque, known as the 'black box effect' of AI processing (Castelvecchi, 2016). Although this learning process is highly efficient and can quickly adapt to vast amounts of data, it cannot provide the reasons and logic behind generated outputs, leading to potential errors, such as misidentifying an image with orange and black stripes as a school bus. The opacity of the black box hinders people from understanding AI, resulting in a lack of trust in its operations.

To address the problems posed by the black box effect, a possible solution is to employ explainable artificial intelligence (XAI) to generate explanations for AI recognition systems that elucidate the causes of the activity leading to a given classification (Das et al., 2023). For end users, understanding how AI systems reach conclusions is of paramount importance. Explainable AI provides users with insights into the decision-making process, enhancing the user-friendliness of the technology and enabling individuals to make informed judgements based on AI-generated outputs, thus reducing the negative psychological impact of AI (Castelvecchi, 2016). When system users clearly understand the factors influencing a specific output and the scenarios in which the system may not function effectively, they are more likely to trust the decisions (Shin, 2021).

6.5.3 Industry regulations

In the rapidly evolving landscape of the AI industry, establishing industry standards and regulations is crucial for cultivating a healthy environment of competition and collaboration. While legislation usually lags far behind the development of science and technology, industry self-regulation is a pivotal approach to guide the development of AI along ethical and responsible pathways.

Ethical considerations and responsible AI practices should be inherent in the industry. Regulations must provide a framework for delineating ethical standards in both the development and application of AI, thereby setting clear guidelines for AI practitioners and organisations. This will ensure accountability and prevent the misuse of AI technologies that could lead to ethical violations. AI developers should protect users' tangible and intangible assets and recognise and address the psychological impact on individuals. In addition, industry standards should be created to address the potential biases in AI algorithms by mandating fairness assessments and providing guidelines for mitigating biases. This will ensure that AI systems do not perpetuate discrimination or exacerbate existing social inequalities.

Furthermore, international regulation cooperation becomes imperative as AI technologies transcend national boundaries. Global harmonisation of regulations builds trust in AI systems and facilitates the responsible use of AI technologies globally. This collaborative effort underscores the shared responsibility of the global AI community to steer the trajectory of AI development towards ethical and socially responsible practices.

In recent years, an increasing number of industry standards and regulations have been proposed, focusing on the trustworthiness of AI. In April 2019, the European Commission's High-Level Expert Group on Artificial Intelligence presented Ethics Guidelines for Trustworthy Artificial Intelligence, outlining that trustworthy AI should be lawful, ethical, and robust (European Commission, 2019). The International Organization for Standardization (ISO) published an overview of trustworthiness in artificial intelligence in 2020. The survey covered topics such as establishing trust in AI systems, addressing engineering pitfalls and associated threats, and assessing factors like availability, resilience, reliability, accuracy, safety, security, and privacy (International Standards Organisation, 2020). Similarly, the US Government Accountability Office (GAO) published an AI accountability framework to ensure accountability and responsible use of artificial intelligence (AI) in government programmes and processes (US Government Accountability Office, 2021). In June 2023, the IEEE Standards Association approved a standard for security and trustworthiness requirements in Generative Pretrained AI Models. The standard aims to enhance the reliability and trustworthiness of generative pre-trained AI models throughout their entire life cycle, covering development, deployment, and usage (European Union, 2021).

6.5.4 Legal frameworks

Legal frameworks play a crucial role in safeguarding individual rights and privacy in the context of advancing AI technologies. Regulations can set boundaries on the collection, use, and storage of personal data, ensuring that AI systems operate within ethical and legal constraints, thus fostering trust among users.

Legal frameworks can provide a basis for establishing accountability and assigning responsibility for the actions and decisions made by AI systems. In order to enhance control over the potential risks associated with AI systems, many countries and regions are actively working towards developing responsible AI practices (Greenleaf, 2019). In addition, more than 130 countries worldwide have enacted privacy laws and regulations to protect personal privacy (Wachter, 2019). This is essential to ensure that individuals or entities developing, deploying, or using AI technologies are held accountable for any adverse consequences or risks associated with AI applications.

However, we should also recognise that many existing laws concentrate on the unlawful collection or disclosure of data but provide limited coverage for the improper use of data post-collection (European Union, 2022). The issue of 'lawful acquisition and subsequent misuse' remains unaddressed, mainly in the context of big data. In addition, while data protection laws legally govern the procedures of collecting, transmitting, processing, and preserving data from data subjects, it is worth noting that data controllers can generate new information through inference that may not be fully safeguarded by existing data protection laws (Wachter & Mittelstadt, 2019).

The EU's proposed AI Act, introduced in April 2021, categorises AI applications into three risk levels, establishing a framework where applications with unacceptable risks would face prohibition, high-risk applications would be subject to specific legal requirements, and applications not explicitly banned or labelled as high-risk would remain largely unregulated (European Union, 2022; Greenleaf, 2019). Nevertheless, some scholars have raised concerns about the merging of 'trustworthiness' with the 'acceptability of risk', highlighting the potential for a misalignment between the perceived level of trust and the actual trustworthiness of the implemented AI (Laux et al., 2022).

6.6 CONCLUSION

In an era marked by the growing reliance on artificial intelligence, individuals are increasingly impacted by AI, either directly or indirectly. While reaping the benefits of convenience and efficiency offered by AI, there are also inherent risks. Previous research has predominantly focused on tangible risks associated with AI, such as digital, physical, and political security, which are somewhat measurable. However, the deepening interaction between humans and AI necessitates an exploration of their psychological impact on individuals. When AI is assigned tasks, it is essentially entrusted with trust by people, acknowledging the acceptance of potential risks. The interplay between people's trust and AI's trustworthiness significantly contributes to the adoption of AI and the subsequent risks.

Ideally, when individuals possess a moderate understanding of AI, are aware of its risks, and apply appropriate trust in its applications, there is a healthy use of AI. However, over-trusting AI may lead to misuse, such as neglecting risks, lacking critical thinking, or compromising moral standards for AI abuse. Conversely, under-trusting AI may result in disuse, causing potential inefficiencies, missed opportunities, and reduced competitiveness in a society embracing AI advancements.

The consequences of disuse, misuse, and abuse resulting from the misalignment between people's trust and AI's trustworthiness can lead to psychological impacts, whether individuals use AI directly or indirectly. Non-AI users may experience mass impacts like anxiety and insecurity due to the threat of job replacement. Furthermore, potentially biased algorithms, unfair treatment, fake news, and misleading information generated by AI may contribute to psychological distress and distorted perceptions of reality. Privacy concerns arise from AI training data collection, particularly with facial scanning and surveillance networks, leading to chilling effects, self-censorship, and behavioural changes such as fear of expression and avoidance of public gatherings.

For active AI users, excessive reliance on AI can result in over-dependency and a loss of self-confidence when AI is unavailable. The echo chamber effect may limit exposure to diverse information, hindering personal advancement and reinforcing negative beliefs, leading to mental health issues. Additionally, supernormal stimuli created by AI can overly engage individuals in AI-generated virtual environments, causing detachment from reality and harming social and emotional well-being.

Individuals are also impacted by the malicious use of AI, including deception through AI-generated text, video, audio, and connections from individuals with AI-generated profiles. This malicious use not only leads to digital and financial losses but also induces feelings of betrayal, mistrust, and fear, with detrimental effects on mental health.

The authors advocate for balancing user trust and AI trustworthiness to address these risks. This involves cultivating appropriate risk awareness through training, encompassing an understanding of vulnerabilities and security threats, and the development of critical-thinking skills. Simultaneously, efforts should be directed towards enhancing AI's trustworthiness by increasing transparency, refining industry regulations, and establishing robust legal frameworks.

Although 'artificial intelligence' was only invented a few decades ago, it has fundamentally changed our lives. In today's era of artificial intelligence, no matter whether we trust it or use it, everyone will inevitably be affected by it. The impact of artificial intelligence on human mental health should receive more and more attention. Establishing balanced trust in artificial intelligence, using artificial intelligence wisely, and making full use of its capabilities are the best ways for us to deal with it in the AI era. Only in this way can we fully harness AI's potential benefits while mitigating potential risks, ensuring a future where technology positively contributes to the betterment of humanity.

REFERENCES

Abate, A. F., Bisogni, C., Cascone, L., Castiglione, A., Costabile, G., & Mercuri, I. (2020). Social robot interactions for social engineering: Opportunities and open issues. *2020 IEEE Intl Conf on Dependable, Autonomic and Secure Computing, Intl Conf on Pervasive Intelligence and Computing, Intl Conf on Cloud and Big Data Computing, Intl Conf on Cyber Science and Technology Congress (DASC/PiCom/CBDCom/CyberSciTech).* http://dx.doi.org/10.1109/dasc-picom-cbdcom-cyberscitech49142.2020.00097

Abd-Alrazaq, A., Hassan, A., Abuelezz, I., Ahmed, A., Alzubaidi, M. S., Shah, U., & Househ, M. (2021). Overview of technologies implemented during the first wave of the COVID-19 pandemic: Scoping review. *Journal of Medical Internet Research*, *23*(9), e29136. https://doi.org/10.2196/29136

Ajenaghughrure, I. B., Sousa, S. C., & Lamas, D. R. (2018). The interplay between emotion and trust in technology. *Proceedings of the Second African Conference for Human Computer Interaction: Thriving Communities.* https://doi.org/10.1145/3283458.3283494

Alhaji, B., Prilla, M., & Rausch, A. (2021). Trust dynamics and verbal assurances in human robot physical collaboration. *Frontiers in Artificial Intelligence*, *4*. https://doi.org/10.3389/frai.2021.703504

Anania, E. C., Rice, S., Walters, N. W., Pierce, M., Winter, S. R., & Milner, M. N. (2018). The effects of positive and negative information on consumers' willingness to ride in a driverless vehicle. *Transport Policy*, *72*, 218–224. https://doi.org/10.1016/j.tranpol.2018.04.002

Araujo, T., Helberger, N., Kruikemeier, S., & De Vreese, C. H. (2020). In AI we trust? Perceptions about automated decision-making by artificial intelligence. *AI and Society*, *35*(3), 611–623. https://doi.org/10.1007/s00146-019-00931-w

Bach, T. A., Khan, A., Hallock, H., Beltrão, G., & Sousa, S. (2022). A systematic literature review of user trust in AI-enabled systems: An HCI perspective. *International Journal of Human – Computer Interaction*, 1–16. https://doi.org/10.1080/10447318.2022.2138826

Becker, B. A., Denny, P., Finnie-Ansley, J., Luxton-Reilly, A., Prather, J., & Santos, E. A. (2023). Programming is hard – Or at least it used to be. *Proceedings of the 54th ACM Technical Symposium on Computer Science Education V. 1.* http://dx.doi.org/10.1145/3545945.3569759

Blauth, T. F., Gstrein, O. J., & Zwitter, A. (2022). Artificial intelligence crime: An overview of malicious use and abuse of AI. *IEEE Access*, *10*, 77110–77122. https://doi.org/10.1109/access.2022.3191790

Booth, S., Tompkin, J., Pfister, H., Waldo, J., Gajos, K., & Nagpal, R. (2017). Piggybacking robots: Human-robot over trust in university dormitory security. In *Proceedings of the 2017 ACM/IEEE International Conference on Human-Robot Interaction* (pp. 426–434). https://doi.org/10.1145/2909824.3020211

Brown, R. C., Fischer, T., Goldwich, A. D., Keller, F., Young, R., & Plener, P. L. (2017). #cutting: Non-suicidal self-injury (NSSI) on Instagram. *Psychological Medicine*, *48*(2), 337–346. https://doi.org/10.1017/s0033291717001751

Bundage, M., Avin, S., Clark, J., Toner, H., Eckersley, P., Garfield, B., & Defoe, A. (2018). The malicious use of artificial intelligence: Forecasting, prevention, and mitigation. https://doi.org/10.48550/arXiv.1802.07228

Burhanudin, B. (2022). Examining the effect of service value and reputation on customer loyalty with trust and electronic word of mouth as mediation. *Journal Aplikasi Manajemen*, *20*(3). https://doi.org/10.21776/ub.jam.2022.020.03.05

Caroppo, E., Mazza, M., Sannella, A., Marano, G., Avallone, C., Claro, A. E., & Sani, G. (2021). Will nothing be the same again? Changes in lifestyle during the COVID-19 pandemic and consequences on mental health. *International Journal of Environmental Research and Public Health*, *18*(16), 8433. https://doi.org/10.3390/ijerph18168433

Castelvecchi, C. (2016, October 5). *Can we open the black box of AI?* www.nature.com/news/can-we-open-the-black-box-of-ai-1.20731

Chattopadhyay, A., & Beyene, N. (2023). A survey study of psybersecurity: An emerging topic and research area. In *Lecture Notes in Networks and Systems* (pp. 893–912). https://doi.org/10.1007/978-3-031-28073-3_59

Chen, J., Song, N., Su, Y., Zhao, S., & Zhang, Y. (2022). Learning user sentiment orientation in social networks for sentiment analysis. *Information Sciences*, *616*, 526–538. http://dx.doi.org/10.1016/j.ins.2022.10.135

Chen, X., Liu, C., Li, B., Lu, K., & Song, D. (2017). *Targeted backdoor attacks on deep learning systems using data poisoning* (arXiv preprint). https://doi.org/10.48550/arXiv.1712.0552

Chung-Wa, C. K., Cho, E., & Lee, J. (2020). Can an intelligent personal assistant (IPA) be your friend? Para-friendship development mechanism between ipas and their users. *Computers in Human Behaviour*, *111*, 106412. https://doi.org/10.1016/j.chb.2020.106412

Cinelli, M., De Francisci Morales, G., Galeazzi, A., Quattrociocchi, W., & Starnini, M. (2021). The echo chamber effect on social media. *Proceedings of the National Academy of Sciences*, *118*(9). https://doi.org/10.1073/pnas.2023301118

Cleworth, T. W., Chua, R., Inglis, J. T., & Carpenter, M. G. (2016). Influence of virtual height exposure on postural reactions to support surface translations. *Gait & Posture*, *47*, 96–102. https://doi.org/10.1016/j.gaitpost.2016.04.006

Das, D., Nishimura, Y., Vivek, R. P., Takeda, N., Fish, S. T., Plötz, T., & Chernova, S. (2023). Explainable activity recognition for smart home systems. *ACM Transactions on Interactive Intelligent Systems*, *13*(2), 1–39. https://doi.org/10.1145/3561533

Dastin, J. (2022). Amazon scraps secret AI recruiting tool that showed bias against women. In *Ethics of data and analytics* (pp. 296–299). https://doi.org/10.1201/9781003278290-44

Day, T. (2023). A preliminary investigation of fake peer-reviewed citations and references generated by ChatGPT. *The Professional Geographer*, *75*(6), 1024–1027. https://doi.org/10.1080/00330124.2023.2190373

Ding, S., Tian, Y., Xu, F., Li, Q., & Zhong, S. (2019). Trojan attack on deep generative models in autonomous driving. In *Lecture Notes of the Institute for Computer Sciences, Social Informatics and Telecommunications Engineering* (pp. 299–318). https://doi.org/10.1007/978-3-030-37228-6_15

Dorton, S. L., Harper, S. B., & Neville, K. J. (2022). Adaptations to trust incidents with artificial intelligence. *Proceedings of the Human Factors and Ergonomics Society Annual Meeting*, *66*(1), 95–99. https://doi.org/10.1177/1071181322661146

Dremliuga, R., Prisekina, N., & Yakovenko, A. (2020). New properties of crimes in virtual environments. *Advances in Science, Technology and Engineering Systems Journal*, *5*(6), 1727–1733. https://doi.org/10.25046/aj0506206

Dujmovic, J. (2017, March 30). *What's holding back artificial intelligence? Americans don't trust it.* www.marketwatch.com/story/whats-holding-back-artificial-intelligence-americans-dont-trust-it-2017-03-30

Dunlop, S. M., More, E., & Romer, D. (2011). Where do youth learn about suicides on the internet, and what influence does this have on suicidal ideation? *Journal of Child Psychology and Psychiatry*, *52*(10), 1073–1080. https://doi.org/10.1111/j.1469-7610.2011.02416.x

European Commission. (2019, April 8). *Ethics guidelines for trustworthy AI*. https://digital-strategy.ec.europa.eu/en/library/ethics-guidelines-trustworthy-ai

European Commission, & Content Directorate-General for Communications Networks, and Technology. (2019). *Ethics guidelines for trustworthy AI* (pp. 17–21). EU Publications Office. https://doi.org/10.2759/346720

European Union. (2021, September 9). *Proposal for a regulation of the European Parliament and of the council laying down harmonised rules on artificial intelligence*. https://artificialintelligenceact.eu/the-act/

European Union. (2022). *EU artificial intelligence act | up-to-date developments and analyses of the EU AI Act*. https://artificialintelligenceact.eu/

Fast, N. J., & Schroeder, J. (2020). Power and decision making: New directions for research in the age of artificial intelligence. *Current Opinion in Psychology*, *33*, 172–176. https://doi.org/10.1016/j.copsyc.2019.07.039

Fowler, G. (2023, April 3). *We tested a new ChatGPT-detector for teachers. It flagged an innocent student.* www.washingtonpost.com/technology/2023/04/01/chatgpt-cheating-detection-turnitin/

Franklin, C., & Chattopadhyay, A. (2021). Psybersecurity: A new emerging topic and research area within human security, Part 2. *IEEE Technology Policy and Ethics*, *6*(3), 1–4. https://doi.org/10.1109/ntpe.2021.9778095

Gefen, D., Karahanna, E., & Straub, D. (2003). Trust and TAM in online shopping: An integrated model. *MIS Quarterly*, *27*(1), 51. https://doi.org/10.2307/30036519

Gillath, O., Ai, T., Branicky, M. S., Keshmiri, S., Davison, R. B., & Spaulding, R. (2021). Attachment and trust in artificial intelligence. *Computers in Human Behaviour*, *115*, 106607. https://doi.org/10.1016/j.chb.2020.106607

Gratch, J., & Fast, N. J. (2022). The power to harm: AI assistants pave the way to unethical behaviour. *Current Opinion in Psychology*, *47*, 101382. https://doi.org/10.1016/j.copsyc.2022.101382

Greenleaf, G. (2019). *Global data privacy laws 2019: 132 national laws & many bills* (157 Privacy Laws & Business International Report, pp. 14–18). https://ssrn.com/abstract=3381593

Gulati, S., Sousa, S., & Lamas, D. (2019). Design, development and evaluation of a human-computer trust scale. *Behaviour and Information Technology*, *38*(10), 1004–1015. https://doi.org/10.1080/0144929x.2019.1656779

Haw, C., Hawton, K., Niedzwiedz, C., & Platt, S. (2013). Suicide clusters: A review of risk factors and mechanisms. *Suicide and Life-Threatening Behaviour*, *43*(1), 97–108. http://dx.doi.org/10.1111/j.1943-278x.2012.00130.x

Hawksworth, J., Berriman, R., & Goel, S. (2018). Will robots really steal our jobs? An international analysis of the potential long term impact of automation. *Engineering, Computer Science*. https://api.semanticscholar.org/CorpusID:155263517

Hernandez-Ortega, B., & Ferreira, I. (2021). How smart experiences build service loyalty: The importance of consumer love for smart voice assistants. *Psychology & Marketing*, *38*(7), 1122–1139. https://doi.org/10.1002/mar.21497

Holmes, J. G. (1991). Trust and the appraisal process in close relationships. *Advances in Personal Relationships: A Research Annual*, 2, 57–104. https://api.semanticscholar.org/CorpusID:141382336

Hunter, T. (2023, April 27). *Why you shouldn't tell ChatGPT your secrets*. www.washingtonpost.com/technology/2023/04/27/chatgpt-messages-privacy/

International Standards Organisation. (2020, May). *Information technology artificial intelligence overview of trustworthiness in artificial intelligence*. www.iso.org/standard/77608.html

Ives, T. I. (2016). The consequences of cyber attacks. *Journal of International Affairs*, *70*(1), 175–181. www.jstor.org/stable/90012601

Jain, J. (2021). Artificial intelligence in the cyber security environment. In *Artificial intelligence and data mining approaches in security frameworks* (pp. 101–117). https://doi.org/10.1002/9781119760429.ch6

Jakesch, M., Bhat, A., Buschek, D., Zalmanson, L., & Naaman, M. (2023). Co-writing with opinionated language models affects users' views. In *Proceedings of the 2023 CHI Conference on Human Factors in Computing Systems* (pp. 1–15). https://doi.org/10.1145/3544548.3581196

Kaewkitipong, L., Chen, C., Han, J., & Ractham, P. (2022). Human – computer interaction (HCI) and trust factors for the continuance intention of mobile payment services. *Sustainability*, *14*(21), 14546. https://doi.org/10.3390/su142114546

Kearns, I., & Muir, R. (2019). Data driven policing and public value. *The Police Foundation*, *13*(4). www.bath.ac.uk/publications/data-driven-policing-and-public-value/

Kong, H., Yin, Z., Baruch, Y., & Yuan, Y. (2023). The impact of trust in AI on career sustainability: The role of employee – AI collaboration and protean career orientation. *Journal of Vocational Behaviour*, *146*, 103928. https://doi.org/10.1016/j.jvb.2023.103928

Laux, J., Wachter, S., & Mittelstadt, B. (2022). Trustworthy artificial intelligence and the European union AI act: On the conflation of trustworthiness and the acceptability of risk. *SSRN Electronic Journal*, 1556–5068. https://doi.org/10.2139/ssrn.4230294

Lee, J. D. (2008). Review of a pivotal human factors article: "Humans and automation: Use, misuse, disuse, abuse." *Human Factors: The Journal of the Human Factors and Ergonomics Society*, *50*(3), 404–410. https://doi.org/10.1518/001872008x288547

Lewis, M., Sycara, K., & Walker, P. (2018). The role of trust in human-robot interaction. *Foundations of Trusted Autonomy*, 135–159. https://doi.org/10.1007/978-3-319-64816-3_8

Luxton, D. D., & Watson, E. (2023). Psychological and psychosocial consequences of super disruptive A.I.: Public health implications and recommendations. *Proceedings of the Stanford Existential Risk Conference*. https://doi.org/10.25740/mg941vt9619

Maddison, L. (2023, April 4). *Samsung workers made a major error by using ChatGPT.* www.techradar.com/news/samsung-workers-leaked-company-secrets-by-using-chatgpt

McKnight, D. H., & Chervany, N. L. (2001). Trust and distrust definitions: One bite at a time. In *Trust in cyber-societies* (pp. 27–54). https://doi.org/10.1007/3-540-45547-7_3

Mell, J., Lucas, G., Mozgai, S., & Gratch, J. (2020). The effects of experience on deception in human-agent negotiation. *Journal of Artificial Intelligence Research, 68*, 633–660. https://doi.org/10.1613/jair.1.11924

Melville, K. (2019, August 30). *The insidious rise of deepfake porn videos – And one woman who won't be silenced.* www.abc.net.au/news/2019-08-30/deepfake-revenge-porn-noelle-martin-story-of-image-based-abuse/11437774

Merken, S. (2023, June 26). *New York lawyers sanctioned for using fake ChatGPT cases in legal brief.* www.reuters.com/legal/new-york-lawyers-sanctioned-using-fake-chatgpt-cases-l

Myers, J. (2022, January 5). *Five charts that show what people around the world think about AI.* www.weforum.org/agenda/2022/01/artificial-intelligence-ai-technology-trust-survey/

Noonan, P. (2023, March 30). *A six-month AI pause? No, longer is needed.* www.wsj.com/articles/a-six-month-ai-pause-no-longer-is-needed-civilization-danger-chat-gpt-chatbot-internet-big-tech-4b66da6e

Noy, S., & Zhang, W. (2023). Experimental evidence on the productivity effects of generative artificial intelligence. *SSRN Electronic Journal.* http://dx.doi.org/10.2139/ssrn.4375283

O'Brien, M. (2023, April 30). *Musk, scientists call for halt to AI race sparked by ChatGPT.* https://apnews.com/article/artificial-intelligence-chatgpt-risks-petition-elon-musk-steve-wozniak-534f0298d6304687ed080a5119a69962

Odgers, C. L., & Jensen, M. R. (2020). Annual research review: Adolescent mental health in the digital age: Facts, fears, and future directions. *Journal of Child Psychology and Psychiatry, 61*(3), 336–348. https://doi.org/10.1111/jcpp.13190

Parasuraman, R., & Riley, V. (1997). Humans and automation: Use, misuse, disuse, abuse. *Human Factors: The Journal of the Human Factors and Ergonomics Society, 39*(2), 230–253. https://doi.org/10.1518/001872097778543886

Patrini, G. (2019, November 29). *Mapping the deepfake landscape.* https://medium.com/sensity/mapping-the-deepfake-landscape-27cb809e98bc

Perrigo, B. (2023, February 17). *Bing's AI is threatening users. That's no laughing matter.* https://time.com/6256529/bing-openai-chatgpt-danger-alignment/

Pinney, J., Carroll, F., & Newbury, P. (2022). Human-robot interaction: The impact of robotic aesthetics on anticipated human trust. *PeerJ Computer Science, 8*, e837. https://doi.org/10.7717/peerj-cs.837

Raats, K., Fors, V., & Pink, S. (2020). Trusting autonomous vehicles: An interdisciplinary approach. *Transportation Research Interdisciplinary Perspectives, 7*, 100201. https://doi.org/10.1016/j.trip.2020.100201

Raphael, S. (2021, May 1). *Experts: Spy used AI-generated face to connect with targets.* https://apnews.com/article/ap-top-news-artificial-intelligence-social-platforms-think-tanks-politics-bc2f19097a4c4fffaa00de6770b8a60d

Robinette, P., Howard, A., & Wagner, A. R. (2017). Conceptualising overtrust in robots: Why do people trust a robot that previously failed? In *Autonomy and artificial intelligence: A threat or savior?* (pp. 129–155). http://dx.doi.org/10.1007/978-3-319-59719-5_6

Roose, K. (2022, October 21). *A.I.-generated art is already transforming creative work.* www.nytimes.com/2022/10/21/technology/ai-generated-art-jobs-dall-e-2.html

Sample, I. (2020, January 13). *What are deepfakes – And how can you spot them?* www.theguardian.com/technology/2020/jan/13/what-are-deepfakes-and-how-can-you-spot-them

Sanders, T., Oleson, K. E., Billings, D. R., Chen, J. Y., & Hancock, P. A. (2011). A model of human-robot trust: Theoretical model development. *Proceedings of the Human Factors and Ergonomics Society Annual Meeting*, *55*(1), 1432–1436. https://doi.org/10.1177/1071181311551298

Schneier, B. (2019, June 13). *AI has made video surveillance automated and terrifying.* https://cyber.harvard.edu/story/2019-06/ai-has-made-video-surveillance-automated-and-terrifying

Schoorman, F. D., Mayer, R. C., & Davis, J. H. (2007). An integrative model of organisational trust: Past, present, and future. *Academy of Management Review*, *32*(2), 344–354. https://doi.org/10.5465/amr.2007.24348410

Sedgwick, R., Epstein, S., Dutta, R., & Ougrin, D. (2019). Social media, internet use and suicide attempts in adolescents. *Current Opinion in Psychiatry*, *32*(6), 534–541. https://doi.org/10.1097/yco.0000000000000547

Shin, D. (2021). The effects of explainability and causability on perception, trust, and acceptance: Implications for explainable AI. *International Journal of Human-Computer Studies*, *146*, 102551. https://doi.org/10.1016/j.ijhcs.2020.102551

Simpson, J. A. (2007). Psychological foundations of trust. *Current Directions in Psychological Science*, *16*(5), 264–268. https://doi.org/10.1111/j.1467-8721.2007.00517.x

Singh, A., Patil, D., Reddy, G. M., & Omkar, S. (2017). Disguised face identification (DFI) with facial keypoints using spatial fusion convolutional network. In *2017 IEEE International Conference on Computer Vision Workshops (ICCVW)* (pp. 1648–1655). http://dx.doi.org/10.1109/iccvw.2017.193

Starr, A., Fernandez, L. A., Amster, R., Wood, L. J., & Caro, M. J. (2008). The impacts of state surveillance on political assembly and association: A socio-legal analysis. *Qualitative Sociology*, *31*(3), 251–270. http://dx.doi.org/10.1007/s11133-008-9107-z

US Government Accountability Office. (2021, June 30). *Artificial intelligence: An accountability framework for federal agencies and other entities.* www.gao.gov/products/gao-21-519sp

Vosoughi, S., Roy, D., & Aral, S. (2018, March 9). *The spread of true and false news online.* www.science.org/doi/10.1126/science.aap9559

Wachter, S. (2019). Data protection in the age of big data. *SSRN Electronic Journal*, 1556–5068. https://doi.org/10.2139/ssrn.3355444

Wachter, S., & Mittelstadt, B. (2019). A right to reasonable inferences: Re-thinking data protection law in the age of big data and AI. *Columbia Business Law Review*, 2, 494. http://dx.doi.org/10.31228/osf.io/mu2kf

Wang, H., Wang, Q., & Hu, F. (2019). Are you afraid of heights and suitable for working at height? *Biomedical Signal Processing and Control*, *52*, 23–31. https://doi.org/10.1016/j.bspc.2019.03.011

Wang, Y., & Chaudhuri, K. (2018). *Data poisoning attacks against online learning* (arXiv preprint). https://doi.org/10.48550/arXiv.1808.08994

Wells, W. D. (1975). Psychographics: A critical review. *Journal of Marketing Research*, *12*(2), 196–213. http://dx.doi.org/10.1177/002224377501200210

Wiencierz, C., & Lünich, M. (2020). Trust in open data applications through transparency. *New Media and Society*, *24*(8), 1751–1770. https://doi.org/10.1177/1461444820979708

Williams, T. (2023, July 11). *AI text detectors 'biased against non-native English speakers'*. www.timeshighereducation.com/news/ai-text-detectors-biased-against-non-nativ

World Economic Forum. (2020, October 20). *The future jobs report*. www.weforum.org/publications/the-future-of-jobs-report-2020/in-full/

Xiang, C. (2022, September 26). *AI is probably using your images and it's not easy to opt out*. www.vice.com/en/article/3ad58k/ai-is-probably-using-your-images-and-its-not-easy-to-opt-out

Yu, S., & Carroll, F. (2023a). A balance of power: Exploring the opportunities and challenges of AI for a nation. In *Advanced sciences and technologies for security applications* (pp. 15–37). http://dx.doi.org/10.1007/978-3-031-40118-3_2

Yu, S., & Carroll, F. (2023b). Securing privacy during a world health emergency: Exploring how to create a balance between the need to save the world and people's right to privacy. In *Data protection in a post-pandemic society* (pp. 145–167). https://doi.org/10.1007/978-3-031-34006-2_5

Yudkowsky, E. (2023, March 29). *The open letter on AI doesn't go far enough*. https://time.com/6266923/ai-eliezer-yudkowsky-open-letter-not-enough/

Zenger, J., & Folkman, J. (2019, February 5). *The three elements of trust*. https://hbr.org/2019/02/the-3-elements-of-trust

Chapter 7

Security through influence over mandate

Josh Armitage and Oliver Guidetti

7.1 INTRODUCTION

7.1.1 Make security tangible, not another compliance slideshow

Security must be more than a checklist or a set of abstract principles presented in compliance meetings. It should be embedded in the daily workflows and processes, making it a tangible part of the organisational culture. This means creating real-world scenarios and simulations that mirror the organisation's potential security threats and challenges. By doing so, employees can understand the practical implications of security breaches and the importance of adhering to security protocols. Interactive training sessions, gamified security challenges, and hands-on workshops can make security concepts more relatable and engaging, ensuring they are taken seriously across all levels of the organisation.

7.1.2 Ease, not pain, should be synonymous with security

Often, security measures are seen as cumbersome, slowing down operations and hindering productivity. This perception needs to change. Security protocols should be designed with the user experience in mind, ensuring they are seamless and non-intrusive. This can be achieved by integrating user-friendly security tools, automating repetitive security tasks, and adopting single sign-on (SSO) systems to reduce password fatigue. By making security measures easy and convenient, employees are more likely to adopt and adhere to them, enhancing the organisation's overall security posture.

7.1.3 Resistance is necessary and good

Resistance in this context refers to the organisation's resilience to security threats and ability to adapt to new challenges. It is about building a culture where continuously questioning and challenging the *status quo* is encouraged to improve security measures. This involves regular security audits, penetration testing, and encouraging a culture of open feedback where employees feel comfortable reporting potential security issues. By fostering

 DOI: 10.1201/9781032664859-7

an environment where resistance is seen as a means to strengthen security, organisations can become more agile and responsive to emerging threats.

7.1.4 Security is achieved through continuous, small steps

Implementing robust security measures is not a one-time event but a continuous process. It involves regular updates, patches, and improvements to security protocols. This pattern emphasises the importance of taking incremental steps to enhance security rather than attempting to overhaul systems in one go, which can be overwhelming and disruptive. Continuous integration and continuous deployment (CI/CD) practices can be adapted to include security checks and balances, ensuring that security considerations are integrated into every stage of the software development life cycle.

7.1.5 Security is enablement, not gatekeeping

Traditionally, security has been viewed as a barrier to innovation and speed. However, this perspective must shift towards seeing security as an enabler of business agility and growth. Integrating security into the early stages of project development and decision-making processes can facilitate faster time-to-market and innovation. This requires a collaborative approach where security teams work closely with other departments to understand their needs and tailor security measures that support their objectives rather than hinder them. By embracing these patterns, organisations can foster a security culture that is proactive, integrated, and aligned with their business goals. This protects the organisation from potential threats and enhances its competitiveness in the digital age.

7.2 PATTERN 1: MAKE SECURITY TANGIBLE, NOT ANOTHER COMPLIANCE SLIDESHOW

Companies often say security is critical, yet their actions tell a different story. This gap is evident in the widespread use of generic, compliance-focused training modules that fail to engage or educate employees meaningfully. The use of monotonous, artificial voices in training materials, especially in the era of advanced generative AI (GenAI), further alienates employees, making the experience even less engaging. Most security training is poorly crafted compliance slideshows, with poorly generated voices that sound all the worse in the world of GenAI. The measure of success is the compliance percentage, not any visceral security improvement. The conversations off the back of the training come in two flavours: the irritation at slogging through the slides and the competition around how fast you can complete it. When security is genuinely seen as critical, to quote the suffragettes, you see it in deeds, not words. When elevating security across organisations, you reap what you sew.

7.2.1 Adding to the cultural fabric

Actual behavioural change in cybersecurity is achievable when training transcends traditional methods and becomes an intrinsic part of organisational life. Investing time and resources into practical, hands-on activities that resonate with employees can significantly enhance an organisation's security posture far beyond the superficiality of 'vanity metrics'.

By working in a consultancy where time is money, I have seen tangible security training drive the business forward. Routinely, we stopped all client work, and therefore all revenue, for an entire day to attack simulations that made the security impacts of our decisions real and tangible. The stories from these events have since become the myths on which the security culture is founded. Slides cannot shape culture, but behaviour change does. The typical security training standards used across businesses cannot drive real change.

Practical security training changes behaviour and culture; success cannot be measured in a single metric. If taking people away from their day jobs to invest in security is infeasible, then security is not a proper priority. When it is a priority, you can see the effects, such as Amazon Web Services' track record and their oft-repeated quote, 'Security is job zero' (Rodgers, 2018).

Although it often feels like there is no time in the day for security or other supposed priorities, when looking at healthcare, we can see how, even in emergency departments, it is possible to make time for the things that truly matter. In Feel-Good Productivity, Abdaal (2023) talks through his first emergency department shift, where, although the emergency department is overloaded, he is told to ensure he takes a break; otherwise, he risks making avoidable mistakes by pushing through. As you will see in pattern 4, the same applies to security. If we do not make the time, risks compound until a cataclysm eventuates. The right approach is more impactful than the typical playbook we are generally subjected to; a single game day, while seemingly more expensive, will deliver more change than a million slides.

7.2.2 Accelerating the spread of knowledge

In the intricate dance of cybersecurity, the weak link phenomenon is not just a possibility but an inevitability. The stark reality is that a single oversight can unfurl into a full-blown breach, spotlighting the critical need for a uniformly fortified team. Therefore, when one person can inadvertently be the source of the breach, how do you ensure everyone levels up through security exercises?

The crux lies in the human inclination to defer to a central figure during upheavals. This single-person dependency, often latent and at the worst possible moment, emerges in the throes of a crisis, exposing vulnerabilities at the most inopportune juncture.

The imperative is to equip everyone with the acumen to navigate security incidents adeptly. Drawing inspiration from the collaborative ethos of pair programming, one can orchestrate scenarios where knowledge dissemination is encouraged and necessitated. Deliberately assigning seasoned personnel to observer roles sets the stage for the less experienced to engage actively, thereby democratising expertise. By borrowing from the school of pair programming, there are methods where you can channel expertise to uplift less experienced team members. When running security training, by locking certain people into a hands-off role, you force them to let others get hands-on, maximising the learning opportunity. For cyber resilience, institutional knowledge is far more potent than tacit knowledge. A robust cyber posture cannot be built from heroism; it is constructed from ever-raising standards.

The cornerstone of cyber resilience lies in the organisation's collective wisdom rather than isolated champions' prowess. Heroic acts, though commendable, are ephemeral in their impact. In contrast, a culture of continuous improvement and shared learning lays the groundwork for a resilient cyber posture. Therefore, you must ensure everyone's cyber competence is continually raised. While not everyone can or should be experts, your cyber posture is not defined by your peaks but by your troughs. The better you spread cyber knowledge, the more resilient the organisation is. The goal is not to mould every member into a cybersecurity virtuoso but to elevate the baseline understanding and preparedness across the organisation. The vulnerabilities, not the strengths, delineate the boundaries of your cyber defence. By broadening the dissemination of cyber knowledge, you fortify the organisation's resilience.

In times of crisis, people tend to cluster around a select few individuals. However, this instinctual behaviour can be counterbalanced by empowering others with the necessary skills and confidence. Creating an environment where competence and confidence are distributed equally is essential, so everyone is prepared and ready to act when a crisis occurs. The focus should not be on relying on a few individuals but on the collective readiness of the entire group.

Team of Teams by McChrystal et al. (2015) argues for optimising effectiveness rather than efficiency in complex environments. By spreading security knowledge through the organisation, you build the resilience and adaptability needed to perform in a crisis. It is the same reasoning behind first aid courses, the FAST acronym for identifying strokes and DRSABCD for choking. Not everyone can be a medical expert, but lives are saved daily through a distributed level of competence.

7.2.3 The difference between reading and knowing the incident response plan

Acknowledging that just because procedures like an incident response plan are read does not necessarily mean they are fully understood and retained. The efficacy of traditional methods of familiarising oneself with such critical procedures is often questionable. It is crucial to explore alternative means of learning that may be more effective in retaining information. Typically, the engagement with such a plan is limited to obligatory annual readings, which, while ensuring nominal familiarity, fall short of ingraining the details and protocols deeply into the organisation's collective consciousness. The challenge lies in ensuring that individuals have read the document and in verifying that they comprehend its contents and can recall and apply this knowledge when necessary.

The stark reality is that, despite having ostensibly read the incident response plan, many might struggle to recount its specifics or even pinpoint its location when pressed. This superficial engagement underscores the gap between mere exposure to information and meaningful internalisation.

Transitioning from passive reading to active practice significantly enhances retention and understanding. Engaging in practical exercises that simulate potential incidents can elevate retention rates dramatically, from the paltry 10% associated with reading to a more robust 75%. Furthermore, taking on the role of an instructor and imparting this knowledge to others can further solidify one's grasp of the material, potentially boosting retention to an impressive 90%, as Hallam et al. (1955) suggested.

Simulating realistic scenarios ensures that the theoretical knowledge of the incident response plan imparted is tested and reinforced, making it second nature to the team members. This hands-on approach ensures that, even though 'no plan survives contact with the enemy', the core principles and steps of the plan are so well internalised that they serve as a reliable guide under duress. The plan, thus internalised, transforms into a mental checklist that guides individuals through the desired actions even in high-pressure situations.

It is crucial, therefore, that familiarity with the incident response plan is cultivated well in advance of any actual crisis – regular drills. As explored in 'The Checklist Manifesto' (Gawande, 2014), checklists improve outcomes under pressure, irrespective of anyone's expertise. They are a cornerstone of the safety culture in both aviation and medicine. Nevertheless, given that it should be the cornerstone of our incident response, the plan rests collecting dust on a virtual shelf.

The ability to respond to an incident is a matter of experience, often acquired by dealing with unforeseen situations. Incidents are the most expensive way to gather said experience. However, incidents are unpredictable and expensive, primarily when they damage or harm people or property. Simply reading the plan once a year is insufficient to ensure that people retain the information or take appropriate action when needed. To be able to act in a planned and effective manner, people must not only read the plan but also internalise it and make it a part of their daily routine. This way, the plan becomes a living document that they can use to guide their actions and decisions, rather than a static document that is rarely consulted.

Being able to respond to an incident is a matter of experience. Incidents are the most expensive way to gather said experience. If all people do is read the plan annually, you cannot expect them to retain the information, let alone act on it. If you want people to act in a planned fashion, they do not need to read the plan; they need to live it.

7.2.4 The discipline of security chaos engineering

Security chaos engineering is a technical practice that prepares organisations for inevitable cyber incidents (Shortridge & Rinehart, 2023). It involves creating and running realistic security scenarios that simulate potential attacks. By adopting a proactive approach, security chaos engineering can help organisations identify system vulnerabilities and improve incident response. This emerging field is based on Netflix's chaos engineering approach but has been adapted to prioritise cyber resilience. Security chaos engineering recognises that our systems today are socio-technical and that vulnerabilities can arise from both security's technical and social aspects. Unlike traditional techniques like penetration testing and threat modelling that only focus on the technical side, security chaos engineering takes a holistic view. It assumes vulnerabilities exist and exercises teams' incident response to build true cyber resiliency. Security chaos engineering can help organisations prepare for cyber incidents and minimise their impact by exploring the social side of incidents. Security chaos engineering is required for modern organisations, as cyber incidents are inevitable. It is impossible to prove that a system has no vulnerabilities, so organisations must assume that it is a matter of when they will be successfully attacked, not if. The effectiveness of incident response is determined by people more than any process or tool. As all our systems are socio-technical, security chaos engineering effectively validates the security posture of an organisation's systems. By including the human factor in security testing, organisations can improve their incident response and minimise the impact of cyber incidents.

7.2.5 Summary

- The array of slides in front of us as 'security training' does not work.
- Companies undermine their security by taking actions incongruent with their rhetoric.
- Valuable security training becomes part of the organisation's fabric, ensuring the lessons are remembered and embedded.

- Ensure that security training optimises knowledge sharing, not building single-person dependencies.
- If all people ever do is read, you cannot expect them to act as you intend during a crisis.
- Security chaos engineering is an emerging practice that embodies this sentiment and is built for the reality of socio-technical systems.
- You need to make security real and tangible to expect people to care genuinely.

7.3 PATTERN 2: EASE, NOT PAIN, SHOULD BE SYNONYMOUS WITH SECURITY

There is a widely accepted belief that making something more secure often makes it harder to use. However, if we make something too complex when trying to change behaviour, we discourage such change. Like lightning, humans tend to seek the path of least resistance. Therefore, we must ensure that the secure option is also the easiest. Unfortunately, this aspect is often overlooked when implementing new security measures. Making something more complex when trying to change behaviour disincentivises the change. Humans are like lightning; they seek the path of least resistance. We need to make the secure option also the easiest option. Unfortunately, this point is often missed when implementing new security controls. Make something too complicated, and people will find a workaround, undermining everything you try to achieve. It is essential to influence people's behaviour towards security rather than mandating it.

7.3.1 Mindfully managing tension

Cybersecurity is a delicate balancing act that requires managing the needs and requirements of different groups with competing interests. As a cybersecurity professional, your primary responsibility is maintaining security while ensuring operations run smoothly. However, it is equally important to understand and empathise with the people affected by your decisions. Failure to do so can lead to significant resistance and pushback.

Incentivising compliance is essential to strike the right balance instead of punishing non-compliance. As Princess Leia Organa famously said, 'The more you tighten your grip, the more star systems will slip through your fingers'. When implementing security measures, you have two options to reduce the burden on people: abstract the control or provide training to make it easier to observe. Abstraction involves simplifying the process using fingerprint logins instead of complex password schemes. On the other hand, training is about helping people understand the importance of security and handling sensitive data correctly. It is not just about the control itself but also about the cognitive load the control places on people (Sweller, 1988). By understanding this delicate balance, you can ensure that your cybersecurity measures are effective and that your organisation's operations run smoothly.

You need to avoid having blinkers on when looking at security controls; there is not just a risk impact of introducing a control; there is an operational one as well. Managing this tension between mitigating risk and impeding operations is crucial to long-term success. As Sisyphus would tell you, you can only push uphill for so long (Fry, 2018). Every control is a trade-off between risk and operations; you can lessen the operational impact through abstraction and training, building momentum within your company instead of resistance.

7.3.2 Dealing with a complex world

A robust protective posture is the central goal of cybersecurity. This requires a strategic approach that goes beyond merely implementing a large number of controls. Unfortunately, many organisations adopt a shallow approach that solely relies on adding controls without considering their value. As a result, the cybersecurity landscape becomes cluttered with redundant, ineffective, or even counterproductive controls. Such an approach can make it challenging to identify and address the real threats and can ultimately lead to a false sense of security.

The Cynefin framework shows how you must understand your domain before taking appropriate action. The organisation you work for and its technical systems are complex adaptive systems (Miller & Page, 2009; Snowden, 2020). Therefore, you are working within the complex domain. This means you cannot predict the consequences of an action before you take it. It would be best to verify that the controls you are implementing are causing the correct behaviour changes and do not have unintended side effects that outweigh the value of the control.

It is essential to ensure that control has the desired impact when implemented. Many controls are put in place despite not providing much value. For example, password rotation policies are ineffective but still used in many places. In complex systems, it is hard to predict cause and effect. So, when introducing a new control, it is essential to carefully examine its impact and ensure it causes the intended changes in behaviour without any adverse effects. Since these systems are complex, what works in one situation might not work in another. That is why having a tailored and adaptable approach to cybersecurity is essential.

One crucial aspect of cybersecurity that is often overlooked is the need to thoroughly assess the impact of control after it has been implemented. A classic example of this is the practice of password rotation policies, which are widely adopted but may not always be effective and can even lead to user fatigue and insecure password practices. Verifying that controls have the desired impact and eliminating ineffective ones is essential. This requires the ability to choose and implement controls and the discernment to identify and remove those that fail to add value. By continuously evaluating and adjusting the cybersecurity framework, it can remain agile, relevant, and effective without being weighed down by outdated or ineffective measures.

7.3.3 Beware premature abstractions

While abstractions can be valuable, creating a helpful abstraction is tricky; for example, Hohpe (2020) emphasises the inherent value of abstractions while cautioning against the complexities of creating constructive ones. Abstracting a problem is not merely an act of simplification but requires a profound understanding of the underlying issues. This understanding is often overlooked, leading to premature abstractions that bring more harm than good, causing frustration and inconvenience for those interacting with these abstractions, particularly in contexts as demanding as developer workstations.

Developers, known for their rigorous requirements, often find themselves at odds with overly simplified solutions that aim to provide security through an 'out-of-the-box' experience. These solutions, while well-intentioned, fail to encompass the myriad of unique and intricate needs that developers face. The challenge lies not in creating a one-size-fits-all abstraction but in empowering developers to tailor solutions within a framework of guidelines. This approach acknowledges the limitations of abstractions and the importance of providing flexibility and ownership to the end users.

Recognising the limitations of one's abstractions is essential in developing effective solutions. Abstractions can be powerful tools that alleviate problems when executed with precision and care. However, when done hastily or without sufficient understanding, they can exacerbate issues, eroding trust within an organisation. Each unsuccessful abstraction compounds the difficulty of gaining acceptance for future endeavours, underscoring the necessity of a cautious and informed approach.

When dealing with cybersecurity, it is essential to use abstraction cautiously. Although it can help simplify complex issues and make them more manageable, it is essential to consider the particular requirements and challenges of the situation before applying it. Only then can we fully leverage its potential to enhance security measures significantly. The key lies in striking the right balance between simplification and customisation, ensuring that abstractions empower rather than hinder the individuals they are designed to assist.

7.3.4 Security is dynamic, not static

It is essential to understand that security is a dynamic system that is not static. This aligns with the business adage that change is the only constant, highlighting the need for cybersecurity strategies to be adaptable and forward-thinking. To effectively deal with the ever-evolving landscape of cybersecurity, it is necessary to employ thoughtful methods for decision-making. This ensures that security measures are robust and responsive to emerging threats and vulnerabilities.

The goal of enhancing a business's security posture extends beyond merely attempting to halt change; it involves actively shaping it to foster a secure yet flexible environment. This approach is exemplified in managing developer workstations, where the emphasis is on enabling developers to implement rapid changes securely. For instance, when developers require the installation of new software, providing an automated, dynamic risk assessment facilitates their progress while maintaining security. This focus on enabling movement rather than maintaining a static position acknowledges the inherent nature of change and the importance of securing it.

The concept of 'desire paths', as discussed by Tommes (2022), serves as a poignant metaphor in this context. In urban planning, desire paths emerge as individuals naturally seek the most efficient routes, often diverging from the meticulously designed pathways. Rather than attempting to restrict access to these user-created paths, formalising them is more pragmatic and beneficial, thereby ensuring their safety and utility. This analogy aptly illustrates the necessity of being attuned to the dynamic nature of systems, including cybersecurity, and the value of adapting to rather than resisting natural tendencies and behaviours.

Viewing security posture as a dynamic system is crucial for accurately reflecting the fluid nature of cybersecurity threats and solutions. It is not about controlling the security posture at any given moment but influencing its trajectory over time. This perspective encourages a focus on shaping positive changes and trends in security practices rather than fixating on static metrics such as the number of control violations or findings from penetration tests.

The concept of living in the 'first derivative', as articulated by Hohpe (2020), emphasises the importance of focusing on the rate of change and trends in cybersecurity rather than absolute positions at any given time. By adopting this mindset, organisations can navigate the complexities of cybersecurity more effectively, ensuring that their strategies are responsive to current challenges and anticipate future developments, focusing on trends, not specific numbers.

7.3.5 Take the product mindset

Adopting a customer-centric product mindset when delivering security solutions significantly shifts from traditional project-based approaches. This perspective focuses on creating compelling security solutions that meet users' needs and encourage their adoption by being inherently desirable and impactful. The essence of a product mindset lies in its emphasis on providing value and impact, aiming to align security measures with user convenience, challenging the conventional dichotomy that positions security and ease of use as mutually exclusive.

Historically, security has often been framed within an adversarial context, emphasising restrictions and limitations instead of enabling and empowering users. This approach has frequently led to resistance and a lack of progress in enhancing organisational security. On the other hand, when security solutions are designed to be appealing and to integrate seamlessly with users' workflows, they gain traction and facilitate widespread organisational adoption. This concept aligns with Marc Andreessen's idea of product-market fit. The market eagerly adopts and pulls the most successful products rather than being pushed onto users (Griffin, 2024).

Achieving product-market fit with security solutions can fundamentally alter the security landscape within an organisation. It shifts the focus from completing security projects to creating sustainable, user-friendly security products that resonate with end users. This approach emphasises outcomes over outputs, encouraging security teams to take ownership of challenges and prioritise actions that benefit the organisation most.

The principle of treating internal users with the same level of consideration as external customers is crucial in this context. Just as a product approach is employed to meet the needs and preferences of external customers, internal users deserve security solutions designed with their needs, preferences, and conveniences in mind. This philosophy underpins the move towards transformative security practices that are effective and embraced by users across the organisation.

7.4 PATTERN 3: RESISTANCE IS NECESSARY AND REASONABLE

Adopting a customer-centric product mindset, rather than viewing security as a mere series of projects, fundamentally shifts the approach to delivering security solutions within an organisation. A product mindset prioritises the creation of solutions that are not only compelling but also inherently secure, thereby making security an integral, desirable feature rather than a cumbersome add-on. This approach is predicated on the impact of the security measures, aiming to make secure practices both accessible and attractive to users. It challenges the traditional, adversarial view of security, which often pits ease of use against security, creating a barrier to widespread adoption and genuine progress.

Marc Andreessen's concept of product-market fit, as cited by Griffin in 2017, is particularly relevant here. Achieving product-market fit in the context of security means developing solutions that users actively want and need, effectively pulling the product into the market. This alignment transforms the role of security within an organisation, moving it from a mandatory, often resisted requirement to a valued and sought-after feature. By focusing on outcomes rather than outputs, a product mindset empowers security teams to assume ownership of challenges, prioritise effectively, and make a tangible impact. It raises the question of why internal customers – essentially, the organisation's employees – should receive less consideration than external customers in developing and implementing security solutions.

Embracing change and managing resistance is critical to implementing a product-centric approach to security. As Block (2023) suggested in 2023, resistance is inevitable and a sign of meaningful change. Organisations are composed of individuals with established patterns of behaviour, often resistant to alteration, encapsulated in the phrase 'that is how we do things here'. Understanding that both excessive and insufficient resistance can herald failure is crucial. Successfully navigating this resistance, identifying its sources, and employing mitigation strategies are essential for enacting lasting change.

In this context, the ability to manage resistance effectively is directly linked to the strength of an organisation's cybersecurity posture. Implementing change on a large scale, particularly in cybersecurity, requires a nuanced understanding of human behaviour, organisational culture, and the delicate balance between innovation and tradition. By approaching security with a product mindset and adept at managing resistance, organisations can foster a culture that accepts and embraces secure practices, leading to a more resilient and adaptable cybersecurity posture.

7.4.1 People react differently to change

Models help to reason about how change permeates through an organisation. The model introduced by Wardley (2023), categorising organisational members into explorers, villagers, and town planners, offers a valuable framework for understanding how change is adopted and implemented within an organisation. This model suggests that an organisation comprises three broad categories of people and teams, each with distinct attitudes towards change and innovation.

Explorers are characterised by their affinity for novelty and eagerness to venture into uncharted territories. They are the pioneers within an organisation, always looking for new methodologies, technologies, and processes that could potentially enhance its capabilities. Their role is crucial in initiating change, as they are the primary source of innovation and new ideas.

Following the explorers are the villagers. This group is less inclined to seek out discoveries independently but is receptive to adopting and expanding upon the innovations introduced by the explorers. Villagers play a vital role in refining and adapting these new ideas, making them more accessible and practical for broader use within the organisation. Their efforts ensure that the innovations introduced by explorers are not isolated but integrated into the organisation's existing frameworks and practices.

Finally, the town planners are responsible for industrialising and standardising the innovations the villagers have proven and accepted. They focus on scalability and efficiency, ensuring that new practices can be implemented widely across the organisation with minimal friction. The town planners enable the organisation to achieve economies of scale, thereby maximising the benefits of the innovations introduced by the explorers and refined by the villagers.

Organisations can strategically manage change initiatives by recognising these three groups' distinct roles and contributions. Initiatives should ideally commence with the explorers, who are most likely to embrace and champion new ideas. As these ideas become more established and validated, the responsibility can then transition to the villagers, who will adapt and expand upon them for broader application. Ultimately, the town planners will take over, focusing on standardisation and broad implementation, ensuring that the innovations are firmly embedded within the organisation's operations.

This model helps identify the right people to be involved at different stages of an initiative and also understand the natural progression of change within an organisation. By aligning initiatives with the inherent tendencies of each group, organisations can facilitate smoother transitions, reduce resistance, and enhance the overall efficiency and effectiveness of their change management processes.

The Japanese principle of Nemawashi, 'Turning The Roots', is the idea that change needs to be prepared for. In a business context, this can be first identifying and then leveraging a stakeholder map where people are engaged and given the ability to voice concerns. There is a human tendency when blind-sided or challenged to entrench existing positions; you can counteract this through thoughtful and careful preparation, ensuring different parties are aligned ahead of time.

Throughout my years as a consultant, there have been cases where our findings have been quite incendiary, and we needed to take great care to ensure our message was received with appropriate intent. Especially in more prominent organisations, with more scar tissue and greater inertia, the differentiator of success is how the message is delivered, more than the content itself.

7.4.2 The nature of bureaucracy

The perception of bureaucracy as predominantly negative is widespread, yet it is crucial to recognise its role as an emergent characteristic of successful organisations. Bureaucracy, at its core, aims to codify practices that have historically led to success, thereby facilitating scalability and efficiency within an organisation. The contention arises not from bureaucratic systems *per se* but from instances where the original intent of these systems becomes obscured, leaving behind rigid, inflexible rules.

In 'The Delicate Art of Bureaucracy', Schwartz (2020) delineates what constitutes effective bureaucracy, emphasising that it should be lean, enabling, and capable of learning. These qualities, however, are often elusive in many organisations, leading to the widespread frustration associated with bureaucratic processes. Schwartz proposes three innovative strategies for navigating and mitigating the challenges of bureaucracy, each embodying a unique approach to problem-solving within constrained systems:

The Monkey: This approach is characterised by an optimistic belief in possibilities. By operating within the established rules, 'Monkeys' leverage their relationships and open-mindedness to discover creative solutions amidst bureaucratic challenges. Their ability to navigate the system without directly challenging its foundations allows them to achieve their objectives through ingenuity and tact.

The Razor: 'Razors' adopt a more analytical stance, focusing on eliminating unnecessary costs and inefficiencies within bureaucratic processes. They quantify the financial impact of delays and other bureaucratic impediments, using data-driven arguments to advocate for change. This approach often garners support from higher levels of management, as it presents a clear business case for streamlining processes to enhance value creation.

The Sumo Wrestler: This strategy involves confronting bureaucratic obstacles by identifying and leveraging their points of vulnerability. 'Sumo Wrestlers' demonstrate the ineffectiveness or counterproductivity of specific bureaucratic measures, particularly how they may exacerbate risks or inefficiencies, such as extended lead times that increase the vulnerability window for security threats.

These approaches underscore the importance of empathy and understanding in overcoming bureaucratic resistance. One can identify mutually beneficial solutions that address individual and organisational needs by comprehending the motivations, objectives, and incentives of those entrenched in the bureaucratic system. This perspective facilitates the navigation of bureaucratic landscapes and fosters a more collaborative and constructive approach to reforming and improving these systems.

7.4.3 Use data to attack the problem

In organisational environments, pressures often exacerbate divisions, leading to an 'us versus them' mentality that can hinder effective collaboration and problem-solving. A data-driven approach can be instrumental in counteracting this tendency and shifting the focus from personal conflicts to the issues at hand. Leveraging data allows for an objective assessment of situations, facilitating a more constructive dialogue across different factions within an organisation.

When implementing new policies or controls, such as those related to security, it is crucial to understand and quantify their implications comprehensively. For instance, introducing a new security control might enhance protection but could also introduce additional workload or complexity for the affected teams. One can present a nuanced perspective that acknowledges the trade-offs by measuring the induced friction, such as the extra effort required to comply with the new control. This approach enables transparent communication and fosters a culture of integrity, as it demonstrates a commitment to considering the impacts on all parties involved.

The 'in tension' metrics concept is pivotal in this context. It involves developing metrics that capture the dynamic interplay between competing priorities, such as the need to maintain high throughput while ensuring stability and minimising risk. This balanced approach prevents the overemphasis on a single aspect, which could lead to unintended consequences elsewhere in the organisation. For example, excessively focusing on maximising throughput without regard to stability can increase the risk of errors and system failures, negating the benefits of the increased output.

By adopting 'in tension' metrics, organisations can navigate the delicate balance between various objectives, ensuring that efforts to optimise one area do not detrimentally affect others. This holistic perspective promotes adopting sustainable practices that yield net positive outcomes for the organisation. It encourages a collaborative approach to problem-solving, where diverse teams work together to achieve a common goal rather than working at cross-purposes due to narrowly defined or conflicting objectives.

7.4.4 Summary

- Resistance is a sign of change; too much and too little resistance are problematic.
- Using models like Wardley's (2023) explorers, villages, and town planners allows you to engage the right people at the right time.
- The principle of Nemawashi reminds us to mindfully set the groundwork for change before prematurely attempting to enact it.
- Bureaucracy is an inevitable aspect of organisational functioning. Therefore, it is essential to understand how to navigate it. Bureaucracy is an inevitable aspect of organisational functioning. Therefore, it is essential to understand how to navigate it.
- By leveraging data, you can break away from the us versus them divides that pervade organisations. The correct data transcends boundaries, enabling open and productive conversations.

Security is an infinite game; success is continually being able to keep playing. Managing resistance is critical to ensuring that you can.

7.5 PATTERN 4: SECURITY IS ACHIEVED THROUGH CONTINUOUS SMALL STEPS

In organisational security, the ideal scenario often involves ample time to plan and implement security measures meticulously. However, time is scarce, relying solely on hope for security initiatives; as Murphy et al. (2016) cautioned, 'Hope is not a strategy'. In an organisation's complex and dynamic environment, the predictability of outcomes from specific actions is limited, underscoring the need for a more agile and experimental approach to change and security enhancements.

Navigating change through a series of small experiments offers a pragmatic alternative to the traditional, overly optimistic planning methods. This approach entails initiating minor, manageable experiments to test hypotheses related to security improvements. The outcomes of these experiments then inform the following steps: whether to escalate the initiative based on successful results, adjust the strategy if the results are promising but not definitive, or abandon the approach if it proves ineffective. This iterative process facilitates a more responsive and adaptable security strategy, allowing organisations to refine their approaches based on empirical evidence and real-world feedback.

Security initiatives have traditionally been approached with a waterfall methodology, as Royce (2021) outlined, characterised by sequential, non-iterative stages. This method assumes that requirements can be fully defined upfront and that changes will be minimal, an assumption that is frequently at odds with the volatile nature of security threats and organisational dynamics. While structured and straightforward, the waterfall approach often leads to rigidity and challenges adapting to new information or changing circumstances.

In contrast, treating security initiatives as a series of small iterative experiments aligns more closely with agile methodologies predicated on flexibility, continuous improvement, and adaptation. Organisations can develop more resilient and effective security postures by embracing a mindset encouraging experimentation, learning from failures, and rapidly iterating on successes. This shift from large, inflexible projects to a more agile, experimental approach increases the likelihood of identifying effective security solutions. It fosters a culture of continuous learning and adaptability within the organisation.

7.5.1 The flow of work

In the landscape of modern organisations, the concept of value streams plays a pivotal role, serving as the channels through which knowledge workflows, from inception to delivery, encapsulating the journey from 'please' to 'thank you' or from idea to revenue. As delineated by the Business Architecture

Guild (2023), every organisation comprises one or more such value streams, within which security work either contributes to or directly influences the outcomes.

As Kersten (2018) advocates in 'Project to Product', the transition from a project-centric approach introduces the Flow Framework, which offers a nuanced understanding of the dynamics within value streams. A critical element of this framework is the Flow Distribution, a method that categorises work into four distinct classes: feature, defect, debt, and risk. This categorisation helps identify the origin of each type of work – features from product owners, defects from customers, debt from the team, and risk from security and other risk-focused entities – and also facilitates an analysis of how teams prioritise and execute security-related tasks over time.

However, a challenge emerges in cybersecurity within value stream-aligned teams, as highlighted by Skelton and Pais (Skelton & Pais, 2019). The inherent focus on delivering immediate value often results in the marginalisation or deferral of security work, a phenomenon that can be attributed to hyperbolic discounting, a concept described by Grüne-Yanoff (2015); this cognitive bias leads individuals and teams to prioritise short-term gains over long-term benefits, thus relegating crucial security tasks.

Organisations must implement mechanisms that safeguard against the natural inclination to deprioritise security work to mitigate this tendency and ensure a balanced approach to addressing security concerns. One effective strategy involves setting minimum allocations for addressing technical debt and managing risk, ensuring that these critical areas are not overshadowed by the immediate allure of new features or the urgency of fixing defects. This approach of 'weeding and feeding' is essential to maintain the health and security of the organisational ecosystem.

Moreover, postponing tasks to ostensibly less busy future dates often results in a perpetual cycle of deferral, as future calendars appear deceptively open only because upcoming tasks have not yet been scheduled. This scenario underscores the necessity of systems that guarantee the availability of resources for ongoing, incremental progress in recognition of the principle articulated by Parkinson (1955) that 'work expands to fill the time allotted for its completion'. Such systems ensure that security and maintenance efforts are continuously advanced, preventing the need for overwhelming corrective measures down the line.

7.5.2 Delivery constraints are security constraints

The planning and execution of security initiatives often revolve around the capacity of the security team, which overlooks a critical aspect: the bulk of implementation responsibilities fall on teams outside of security. This discrepancy necessitates a shift in planning focus from the security team's capacity to orchestrate initiatives to the capacity of other teams to implement them effectively.

The Tame Flow approach, as presented by Tendon (2022), offers a structured methodology for applying the principles of the Theory of Constraints to the realm of knowledge work, particularly within complex environments characterised by multiple projects, events, stakeholders, and teams (PEST environments). In such settings, understanding and addressing constraints are paramount for achieving security objectives (Goldratt & Cox, 2016).

The Theory of Constraints outlines five focusing steps that serve as a roadmap for continuous improvement in security execution:

1. *Identify the Constraint*: The first step involves pinpointing the bottleneck or constraint that limits the system's throughput. In the security context, this could range from a lack of awareness and training among non-security teams to technical or procedural bottlenecks that hinder the implementation of security measures.
2. *Optimise the Constraint*: Once identified, the next step is to optimise the functioning of the constraint to improve its throughput. This could involve enhancing skills, streamlining processes, or introducing tools that alleviate the bottleneck without overwhelming the rest of the system.
3. *Subordinate to the Constraint*: This step requires aligning the entire system's workflow with the constraint's capacity, ensuring that it is not burdened with more work than it can handle. It is about pacing the work according to the constraint's rhythm to avoid exacerbating bottlenecks.
4. *Elevate the Constraint*: If the constraint continues to limit throughput despite optimisation efforts, this step involves making structural changes to elevate the system's overall capacity. This could mean adding resources, making substantial process changes, or investing in significant technological upgrades.
5. *Iterate*: The final step is to recognise that once a constraint is addressed, another will eventually emerge as the system's throughput increases. The process then iterates, starting from the first step, to identify and address the new constraint.

The initial three steps for security teams are particularly crucial as they can be undertaken internally to improve the flow of security work without requiring direct action from external teams. This involves effectively managing the security work backlog, contingent upon having the correct data to identify constraints. Data might be sourced from agile planning tools like Jira, facilitating informed decisions about what should be included in the backlog and its prioritisation.

Organisations can significantly improve the throughput of security initiatives by adeptly shaping the security work backlog. This approach enhances the efficiency of security teams and ensures that the broader organisation is better positioned to implement necessary security measures, thereby strengthening the overall security posture.

7.5.3 Aligning empowerment with incentives

The challenge of achieving security goals within an organisation is often compounded by the structure of the organisational operating model, particularly in terms of where operational and risk accountability and empowerment are allocated. Commonly, the responsibility for managing risk is placed on the shoulders of security teams, yet the authority and capability to resolve many security issues reside with delivery teams. This misalignment can hinder the prompt and effective resolution of security vulnerabilities.

Winters et al. (2020), in their work *Software Engineering at Google*, explore an alternative approach for managing certain classes of vulnerabilities, which could serve as a model for other organisations. At Google, the onus of service health is on the delivery teams, fostering a culture where those directly involved in service development are also accountable for its security. This model is supported by a shared empowerment structure between delivery and security teams, allowing both to take autonomous actions when necessary. Such an approach is convenient in the context of managing dependency vulnerabilities.

The log4shell vulnerability is a pertinent example, illustrating the efficiency of Google's model (Wortley et al., 2021). With minimal coordination required, Google's ability to swiftly deploy a patched version of the affected library starkly contrasts with the protracted processes observed in many traditional enterprises. Updating a vulnerable dependency often entails extensive coordination across numerous teams, leading to significant delays and operational strain.

Transitioning to this more autonomous and empowered model necessitates considerable technical enablement, including implementing robust automated testing and quality assurance processes. These systems are crucial for maintaining service health, especially when changes are introduced externally or by other teams within the organisation.

Organisations' immediate objective should be to optimise the existing network of value streams, enhancing the efficiency and responsiveness of the current operating model. The ultimate aim, however, should be to evolve towards an operating model that reduces the reliance on extensive human-to-human coordination for executing security measures. This not only streamlines the process of addressing security issues but also aligns with broader organisational goals of agility, efficiency, and resilience in the face of evolving security threats.

7.5.4 Summary

- As you are working in a complex space, security is best achieved through many small experiments and pieces of work.
- Organisations are made up of value streams; if you can understand how work gets done, you can maximise the amount that is done.
- The Flow Distribution allows you to identify the flow of security work amongst all other kinds.
- By applying the theory of constraints, you can shape the work to fit the value streams you have today.
- Over the longer term, the security limits come from your operating model, so you must design one that enables security.

In contrast to the large-scale programmes businesses use today for security, success is found in taking many small steps. One of the critical functions of cybersecurity is making it possible for teams to take those steps continually.

7.6 PATTERN 5: SECURITY IS ENABLEMENT, NOT GATEKEEPING

The evolution of security practices within modern organisations underscores a significant shift from a traditionally restrictive approach to one that prioritises enablement and informed decision-making at every step. This transition is driven by the rapid pace of technological change and the increasing frequency of deployments in contemporary software development environments.

The term 'The Church of No', often used to describe enterprise security departments, reflects the historical posture of security as a gatekeeper positioned squarely on the path to production. When system changes were infrequent, the impact of this gatekeeping role on business operations was minimal. However, in the current landscape, where a significant portion of teams deploy updates multiple times daily or at least weekly, the traditional security model becomes untenable. The demand for agility and speed in development cycles necessitates reevaluating security's role in production.

The transformative power of code in accelerating development processes has led to the realisation that keeping pace with engineering teams requires a deep integration of security principles into the engineering discipline. This evolution began with testing automation, expanded into operations and site reliability engineering, and has now culminated in the emergence of DevSecOps and policy as code. While not every security professional needs to be proficient in coding, understanding and engaging with code has become a crucial competency for security departments.

This paradigm shift from a prevention-centric to an enablement-centric approach in security marks a profound change in how organisations view and implement security measures. Rather than acting as a barrier to production, modern security practices aim to integrate seamlessly with development workflows, providing the tools, policies, and frameworks that enable developers to make secure decisions proactively. This approach facilitates a more dynamic and agile development process and fosters a culture of security awareness and responsibility across all levels of the organisation. By empowering teams to incorporate security considerations into their work from the outset, organisations can achieve a more resilient and secure operational posture without sacrificing the speed and innovation that modern markets demand.

7.6.1 One size does not fit all guardrails

The insights derived from Kurtz and Snowden's (2003) Cynefin framework offer a nuanced understanding of how controls can be categorised within organisations, particularly distinguishing between complex and complicated domains (Figure 7.1). Complicated controls, which are well-understood and can be addressed with expert knowledge, lend themselves to automation,

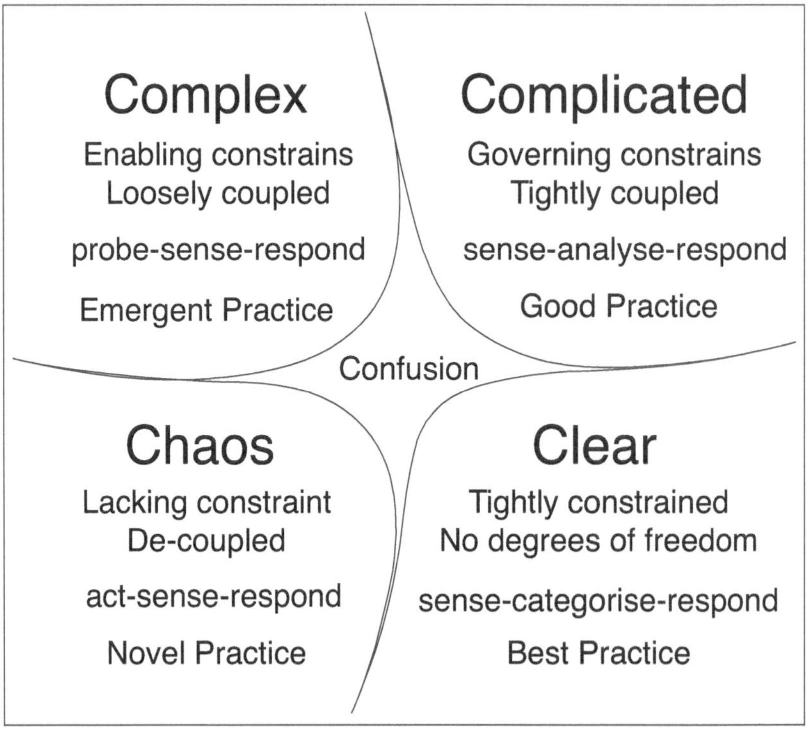

Figure 7.1 Process Plant Operations Certificate II Syllabus Related to IT/OT Brownell (2022).

allowing organisations to establish automated guardrails that pre-empt known issues. On the other hand, complex controls, due to their inherent uncertainty and the interdependent variables involved, necessitate more adaptive approaches, such as threat modelling, to manage the associated risks effectively.

The likelihood of compliance with these controls is significantly enhanced when individuals grasp the underlying reasons for their implementation. Tyler and Balder (2003) emphasise the importance of understanding the rationale behind rules to foster compliance. However, this understanding is often hindered by the curse of knowledge bias, as described by Camerer et al. (1989), where experts may inadvertently assume that everyone shares their specialised knowledge. This assumption can lead to a disconnect between security professionals and engineers, where what seems evident to the former may not be apparent to the latter. Building momentum becomes a Sisyphean task without ensuring everyone understands the why (Fry, 2018). Ensuring a common understanding of the 'why' behind security measures is crucial for building momentum and achieving organisational buy-in.

As detailed in *Team Topologies* by Skelton and Pais (2019), the development and implementation of organisational guardrails typically undergo a natural progression. The authors outline three modes of team interaction: collaboration, facilitation, and as-a-service. Security controls, often by happenstance rather than deliberate design, may fall along this spectrum. The collaborative approach, characterised by security's direct involvement in the production process, is increasingly seen as costly and restrictive. Conversely, the as-a-service model, where policies are imposed rather than co-developed, may lack the engagement necessary for effective implementation.

The transition from collaboration to facilitation and ultimately to an as-a-service model should be managed carefully, especially when new controls are not immediately intuitive to delivery teams. This transition involves creating supporting documentation and reference implementations that facilitate understanding and adoption. It is essential to avoid prolonged stagnation in collaboration or facilitation phases, as this can constrain the scalability and effectiveness of security measures across the organisation.

Moreover, the dynamics of inter-team communication within many organisations tend to develop organically rather than being intentionally structured. This lack of deliberate design, particularly in environments dominated by bureaucratic or pathological cultures instead of generative ones (Westrum, 2004), can lead to entrenched communication patterns that hinder collaborative efforts towards shared objectives. The prevalent us versus them mentality between security teams and other organisational units further exacerbates these challenges, undermining the collective pursuit of enhanced security outcomes. Addressing these cultural and communicative barriers is essential for fostering a more integrated, efficient, and practical approach to organisational security.

7.6.2 Building shared understanding with threat modelling

A deep understanding of the system to be protected is imperative for effectively managing controls in the complex security domain. This goes beyond checkbox approaches and complete automation, necessitating a more nuanced and participatory method like threat modelling. Shostack (2014) outlines threat modelling as a collaborative process to secure a system by identifying potential threats and corresponding countermeasures. This approach addresses security challenges and integrates security considerations into the development process, making security a tangible aspect of system design and implementation.

A significant hurdle in threat modelling is altering the perspectives of various stakeholders regarding the system. Developers often focus on creating value, and security personnel, charged with protecting value, tend to have divergent viewpoints. Bridging this gap is crucial for a cohesive and effective security strategy. The concept of enabling constraints can be instrumental in fostering creativity and innovative problem-solving within defined boundaries. The Elevation of Privilege card game, developed by Shostack, exemplifies this approach by gamifying threat modelling. This innovative tool provides developers with structured prompts that stimulate their ability to identify and mitigate potential security threats, thereby transforming the process from a daunting task into an engaging and enjoyable activity.

As teams become more proficient in threat modelling, it evolves into a self-service function, empowering teams to proactively address security concerns while allowing security professionals to concentrate on strategic objectives. This transition embodies the essence of security as an enabling function, facilitating a proactive and integrated approach to security within the organisation. Simply put, gamification transforms security from a stressor into something playful, potentially something that 'Sparks Joy' (Kondo, 2015).

The contrast in how security is approached in technical versus non-technical domains further highlights the need for a more hands-on, practical approach to security education and enablement. Drawing parallels to the role of a chief fire warden, where practical drills and the distribution of authority are critical components of preparedness, a similar approach could enhance security readiness. Regular, hands-on engagement with security practices, akin to fire drills, could ensure that organisations are better equipped to respond to security incidents. This analogy underscores the importance of treating security with the seriousness it deserves, integrating practical, real-world exercises into the broader security strategy to foster a culture of readiness and resilience.

7.6.3 Summary

- Due to the required pace of change in modern businesses, security needs to focus on enabling teams rather than being the 'Church of No'.
- An engineering capability within security is necessary to scale enablement across an organisation.
- Through Cynefin, you can classify controls into complex and complicated, each requiring different approaches.
- Complicated controls are best turned into guardrails that prevent common security issues from ever occurring.
- Complex controls require a context-specific approach, such as threat modelling.
- By maturing capabilities, such as threat modelling, security can chase higher-order objectives consistently.

Enablement is how security can scale sub-linearly, covering more ground without needing more people. Through engineering and decentralising authority, you can push the organisation forward rather than firefight in perpetuity.

7.7 CONCLUSION

Cybersecurity is pivotal in an era of rapid digital transformation and the critical need for business agility. Esteemed technology behemoths like Google and Amazon exemplify the principle that security forms the cornerstone upon which their rapid innovation and operational speed are built. Despite this, many enterprises persist in the archaic practice of retrofitting security measures onto existing frameworks rather than integrating these measures as fundamental components from the outset.

This chapter has unveiled five quintessential patterns instrumental in reshaping an organisation's security culture, transcending the realms of security and software delivery to offer universal relevance across sectors. These patterns advocate for a reimagined approach to security – transforming it from a mere compliance requirement into a palpable, intrinsic part of the organisational ethos, from a perceived impediment to productivity into a facilitator of seamless and secure operations, and from a rigid gatekeeping entity into a dynamic enabler of innovation and growth.

Exploration of these patterns revealed a compelling narrative: organisations must pivot from passive, compliance-driven security postures to proactive, action-oriented strategies that make security tangible and integral to daily operations. The juxtaposition of security and productivity as mutually exclusive goals was challenged, underscoring the potential for harmonising security with operational efficiency through a product and customer-centric security perspective.

Resistance, often the knee-jerk response to change, was identified as a critical element to be mindfully managed. Leveraging the principles of Nemawashi, navigating bureaucratic landscapes with agility, and employing data-driven strategies emerged as critical tactics for mitigating resistance and effecting sustainable, long-term security enhancements.

The discourse further illuminated the inadequacy of grand, sweeping security overhauls that dominate traditional roadmaps, advocating for a nuanced understanding of the organisational workflow. This understanding enables the tailoring of security initiatives to fit the unique contours of the operating model, ensuring maximum value and efficiency.

Finally, the anachronistic role of security as the proverbial gatekeeper was contested, highlighting the evolving digital landscape's demand for a shift towards proactively enabling teams to uphold security protocols. The chapter critiqued the practice of imposing security controls either too collaboratively or too prescriptively through policy-as-code, advocating for a paradigm where security functions as a catalyst for right action rather than a deterrent against wrong.

In synthesising these insights, the chapter culminates in a compelling vision for organisational security – a future where security is not an afterthought but a keystone of business agility and innovation. By embracing the outlined patterns, organisations can steer through the complex cybersecurity terrain, fostering an ecosystem where security is seamlessly interwoven with the fabric of organisational processes, thereby empowering all members to contribute to a more secure, resilient, and dynamically agile enterprise.

REFERENCES

Abdaal, A. (2023). *Feel-good productivity: How to do more of what matters to you.* Random House.

Beyer, B., Jones, C., Petoff, J., & Murphy, N. R. (2016). Site reliability engineering: How Google runs production systems. O'Reilly Media. http://landing.google.com/sre/book.html.

Blader, S. L., & Tyler, T. R. (2003). A four-component model of procedural justice: Defining the meaning of a "fair" process. *Personality and Social Psychology Bulletin*, *29*(6), 747–758. https://doi.org/10.1177/0146167203029006007

Block, P. (2023). Flawless consulting: A guide to getting your expertise used (4th ed.). Wiley. ISBN 978-1-394-17730-1. https://www.wiley.com/en-us/Flawless+Consulting%3A+A+Guide+to+Getting+Your+Expertise+Used%2C+4th+Edition-p-9781394177301.

Brownell, P. (2022, January 5). The Cynefin framework [framework]. *ideas.the-organization.com.* https://ideas.the-organization.com/decision-making-in-decentralized-systems-d7c990b5a106

Camerer, C., Loewenstein, G., & Weber, M. (1989). The curse of knowledge in economic settings: An experimental analysis. *Journal of Political Economy*, *97*(5), 1232–1254. http://dx.doi.org/10.1086/261651

Fry, S. (2018). *Mythos: The Greek myths retold*. Michael Joseph.

Gawande, A. (2014). *The checklist manifesto: How to get things right*. Penguin Random House India Private. https://doi.org/10.1016/s2155-8256(15)30310-0

Goldratt, E. M., & Cox, J. (2016). *The goal: A process of ongoing improvement*. Routledge.

Griffin, T. (2024, January 24). 12 things about product-market fit. *Andreessen Horowitz*. https://a16z.com/12-things-about-product-market-fit/

Grüne-Yanoff, T. (2015). Models of temporal discounting 1937–2000: An interdisciplinary exchange between economics and psychology. *Science in Context*, *28*(4), 675–713. https://doi.org/10.1017/s0269889715000307

Hallam, B., Miles, H. C., & Dale, E. (1955). Audio visual methods in teaching. *Art Education*, *8*(2), 15. https://doi.org/10.2307/3184288

Hohpe, G. (2020). The software architect elevator. O'Reilly Media, Inc. ISBN 9781492077541. https://www.oreilly.com/library/view/the-software-architect/9781492077534/.

Kersten, M. (2018). Project to product: How to survive and thrive in the age of digital disruption and flow framework (Illustrated ed.). IT Revolution Press. ISBN 978-1942788393.

Kondo, M. (2015). The life-changing magic of tidying up: The Japanese art of decluttering and organizing. Ten Speed Press. ISBN 978-1607747307.

Kurtz, C. F., & Snowden, D. J. (2003). The new dynamics of strategy: Sense-making in a complex and complicated world. *IBM Systems Journal*, *42*(3), 462–483. http://dx.doi.org/10.1147/sj.423.0462

McChrystal, S., Collins, T., Silverman, D., & Fussell, C. (2015). Team of teams: New rules of engagement for a complex world (Illustrated ed.). Portfolio. ISBN 978-1591847489.

Miller, J. H., & Page, S. E. (2009). Complex adaptive systems: An introduction to computational models of social life. Princeton University Press. ISBN 9781400835522.

Parkinson, C. N. (1955, November 19). Parkinson's law. *The Economist*. www.economist.com/news/1955/11/19/parkinsons-law

Rinehart, A., & Shortridge, K. (2020). Security chaos engineering. O'Reilly Media, Inc. ISBN 9781492080343. https://www.oreilly.com/library/view/security-chaos-engineering/9781492080350.

Rodgers, C. (2018, July 9). Security at AWS. *Amazon Web Services*. https://aws.amazon.com/blogs/enterprise-strategy/security-at-aws/

Royce, W. W. (2021). Managing the development of large software systems (1970). In *Ideas that created the future* (pp. 321–332). https://doi.org/10.7551/mitpress/12274.003.0035

Schwartz, M. (2020). The delicate art of bureaucracy: Digital transformation with the Monkey, the Razor, and the Sumo Wrestler (Kindle ed.). IT Revolution Press. ISBN 978-1950508402

Shostack, A. (2014). Threat modeling: Designing for security. Wiley. ISBN 978-1-118-80999-0.

Skelton, M., & Pais, M. (2019). Team topologies: Organizing business and technology teams for fast flow (Illustrated ed.). IT Revolution Press. ISBN 978-1942788812.

Snowden, D. (2022). Cynefin: Weaving sense-making into the fabric of our world (2nd ed.). Cognitive Edge. ISBN 978-1736852033.

Sweller, J. (1988). Cognitive load during problem solving: Effects on learning. *Cognitive Science*, *12*(2), 257–285. https://doi.org/10.1207/s15516709cog1202_4

Tendon, S. (2022). The book of TameFlow: Theory of constraints applied to knowledge-work management. TameFlow Press / TameFlow Consulting Limited. ISBN 978-9995744120.

The Business Architecture Guild. (2023). Guide to the business architecture body of knowledge. Author. https://www.businessarchitectureguild.org/page/002.

Tomes, H. (2022, September 14). The twists and turns of 'desire paths'. *The Spectator*. www.spectator.co.uk/article/the-twists-and-turns-of-desire-paths/

Wardley, S. [@swardley]. (2023, January 26). Explorer, villager, town planner. *Twitter*. https://twitter.com/swardley/status/1618298060076945414/photo/1

Westrum, R. (2004). A typology of organisational cultures. *Quality and Safety in Health Care*, *13*(suppl_2), ii22– ii27. https://doi.org/10.1136/qhc.13.suppl_2.ii22

Winters, T., Manshreck, T., & Wright, H. (2020). Software engineering at Google. O'Reilly Media, Inc. ISBN 9781492082798. https://www.oreilly.com/library/view/software-engineering-at/9781492082781.

Wortley, F., Thompson, C., & Allison, F. (2021, December 9). Log4Shell: RCE 0-day exploit found in log4j, a popular Java logging package. *LunaSec*. https://www.lunasec.io/docs/blog/log4j-zero-day/

Chapter 8

Human skills and sociocultural impacts of K-12 cybersecurity education

Leslie F. Sikos, Marnie McKee, Ahmed Ibrahim, and Nicola F. Johnson

8.1 INTRODUCTION

Ubiquitous computing and the ever-increasing prevalence of computing device use raise cyber safety and security concerns that are challenging to address without bloating existing school curricula or extracurricular training. Considering that even digitally literate students experience high exposure to cyberthreats, teaching cybersecurity from a young age is crucial. While many aspects of cybersecurity require a technical background, human factors such as attitude, contextual awareness, critical thinking, motivation, and trust have also played a substantial role in shaping cybersecure behaviour, which can be established through relevant teaching approaches. There are many different approaches to pass on cybersecurity knowledge, skills, and awareness to K-12 students, to increase safe and secure device usage, responsible social interactions and behaviour online, and to promote cybersecurity career opportunities, including integrating cybersecurity content into existing subjects, in-school studies as collaborative projects (both year-long and as intensives), gamification strategies, out-of-school cyber-focused outreach camps, and the development of educational resource packs. However, international approaches to date have almost all focused on secondary school students. However, little is available for primary school students, whose digital device use is increasing; this is true even at younger ages. Educational initiatives for children must include the professional development of teachers, whose competence in teaching digital technologies is often highly dependent on their confidence and experience. In addition, in many countries' primary schools, computing may not have specialised teachers in the first few years if taught at all from early on.

There are a variety of approaches and opportunities for teachers to develop professional knowledge and confidence. However, a sporadic one-off workshop is unlikely to bring behavioural change in the teachers' own safe and secure digital device usage, let alone become capable of developing ideal behavioural patterns in students. Instead, a range of effective professional learning strategies are needed that can be employed by school leaders and classroom

DOI: 10.1201/9781032664859-8

teachers to support student learning in cybersecurity successfully (Johnson et al., 2022). Teachers must have access to engaging teaching resources and professional learning because users are the weakest link in cyberspace. Hence, the early establishment of appropriate behavioural patterns is critical. In an earlier work (Johnson et al., 2022, p. 17), the authors stated:

> The digital environment is constantly changing as new technology is released for consumers and in the world of work. For this reason, the skills and knowledge required by students *now* will be superseded within a short time. Therefore, cybersecurity curriculum skills and knowledge are required to provide a set of skills to equip students with key skills to manage and protect their data, understand the risks associated with cybersecurity threats and be prepared for ongoing learning. In addition, the senior secondary curriculum has a key role in developing skills and knowledge for students studying to work in ICT industries, including emerging careers in cybersecurity.

The rampant and sophisticated nature of cybercrime underscores the increasing concern over insufficient cybersecurity coverage in schools. Daily, there are reports of individuals being scammed, hacked, or having their identities stolen. Education is a crucial strategy to mitigate or prevent cyber incidents by equipping young people with the knowledge and skills to combat common cyberthreats. Schools have initiated various educational programmes to address issues such as cyberbullying. However, it is imperative to note that cyberbullying is nothing but the digital manifestation of traditional bullying (Olweus, 2012). This digital form of bullying can lead to severe consequences for children, including social withdrawal, low self-esteem, anxiety, self-harming behaviours, and even suicidal thoughts, as noted by Lonergan et al. in 2021.

Furthermore, there is evidence of a correlation between online harassment and behavioural issues within the school environment, as indicated by research from Ybarra et al. in 2007. Addressing the complex nature of cyberbullying effectively requires a comprehensive approach that considers psychological, sociological, institutional, and legal factors. De Luca et al. (2022) emphasise the importance of a multi-perspective approach in understanding the cultural-psycho-socio dynamics of cyberbullying and devising effective interventions.

Not having sufficient cybersecurity coverage in schools is an increasingly growing concern. Cybercrime is rife and increasing in sophistication. People learn about individuals who have been scammed, hacked, or had their identities stolen daily. Education is often viewed as a means to – at least partially – prevent or overcome cyber incidents by helping young people become aware of and able to fight common cyberthreats. To this end, various educational initiatives are implemented by schools. For example, a range of initiatives have been introduced to address cyberbullying, yet cyberbullying is merely bullying in another form (Olweus, 2012), with a range of potential adverse effects on children, such as social withdrawal, low self-esteem, anxiety, self-injurious behaviour, and suicidal thoughts (Lonergan et al., 2021). In addition, studies show linkages between online harassment and school behaviour problems (Ybarra et al., 2007).

Students are not spending time learning to recognise the mindset behind malintent and other cognitive general capabilities crucial for cybersecurity awareness. Instead, bad habits are being established within educational settings as students' digital device use grows. Nevertheless, positive changes in attitude will result in whole school communities identifying cybersecurity's immense benefits: peace of mind, increased user confidence, reputation protection, reduced stress, and ensuring individuals, family, and friends do not lose their hard-earned money to threat actors. Humanity is embedded within cybersecurity, and it needs to be recognised to lead a nation onto a path for overcoming industry workforce shortages and building a cybersecure nation.

Two fundamental aspects of cybersecurity education must be promoted. One surrounds technical competencies; the other covers behavioural competencies and the psychology behind those behaviours. The digital technologies learning area of the Australian curriculum aims to develop knowledge of how to use and apply technologies (ACARA, 2023b). The internationally recognised standard *IEEE 3527.1–2020* for digital intelligence promotes teaching students about digital citizenship, literacy, and readiness (Digital Intelligence, 2021). Such initiatives support statements, for example: 'One of the shortcomings in K-12 education is that students are taught to use various technologies, but they are not introduced to the threats they face while using them' (Javidi & Sheybani, 2018).

One of the main foci for those responsible for students' security and safety in cyberspace is to bring about behavioural change at the individual level, i.e., to apply best practices (cyber hygiene) and use protective technologies effectively. Behavioural competencies need to be built up alongside the technical competencies. Cybersecurity awareness is the funnel for that behavioural change. Hence, a second critical area of focus is on identifying malicious intent. Increasing vigilance and suspicion of criminal deception and strengthening risk assessment are also part of the many behaviours and associated awareness needed to bring about change to mitigate the consequences of inaction.

According to a survey conducted in 2022 among high school students, teachers, and parents in nine European countries (Romania, Slovenia, Greece, Cyprus, Italy, Spain, Germany, the Netherlands, and Poland), the following topics are the most important to be introduced into secondary school curricula (Blažič & Blažič, 2022):

- Being safe in online social platforms
- Recognising fake accounts

- Ensuring privacy in online activities
- Creating strong passwords
- Securely using email applications

As this self-reported data, what is *being emphasised in K-12 curricula* to bring about change that includes both behavioural and technical competencies? What *is* evident is that children and young people (not to mention adults) are using digital devices increasingly and from earlier ages compared to previous generations. However, the use of technology does not necessarily, and often not at all, correlate directly to an increase in knowledge and skills. This is mainly because the user interface of modern computing devices is far more user-friendly and accessible than those in previous decades. This often causes overconfidence in device use and, importantly, casual approaches to data protection, privacy, and safe device use, resulting in a misbelief about the actual capabilities of children in these areas.

Acknowledging that no two children have the same digital knowledge background is essential in developing a scheme for cybersecurity education. Children's exposure to digital services at home and with friends is difficult to measure, as is the interest, attention, and knowledge of parents and occasional caregivers, with whom access to devices and digital services is often extremely different from that at home. In such environments, establishing a baseline of measurement for assumed knowledge is highlighted as a particularly challenging task to understand children's collective and individual learning needs.

A strong argument is that students and teachers (as well as parents and extended school communities) need to obtain in-depth knowledge of technology and critical literacy (the questioning and examination of ideas). This is primarily because digital technologies are often used for entertainment and education (particularly since the COVID lockdowns). Following this logic, *behavioural change* is the final objective because behaviour is the source of security *and* unsecured devices, services, and infrastructures.

Being able to use a digital service does not constitute knowing that service. Instead, the security space is at the point where an individual has evaluated information and is responding to it through critical action. There is a requirement to identify the baseline for educational needs based on teachers', parents' and students' use of digital devices and services, according to those users' actions, or lack thereof. By extension, we must also identify that schools at the holistic level require this same skill and knowledge set. Schools are particularly vulnerable to attacks and must first be aware and vigilant to maintain secure learning environments. For this chapter, cybersecurity refers to the safe and secure use of digital technology and services (as a discipline to be taught) rather than focusing on protecting computer systems and networks by professionals, devices, and tools. Both privacy and ethics define the core of cybersecurity.

8.2 INTERNATIONAL CONTEXT ADDRESSING HUMAN FACTORS IN CYBERSECURITY EDUCATION

Several key international organisations (or at least across multiple continents) have made significant headway in addressing cybersecurity education and its inherent human factors. The bigger picture addresses where the international knowledge of children's cybersecurity education is situated within the broader realm for digital citizens or everyday users of digital devices and services.

Cybersecurity education is at an infancy stage, requiring a structuring of knowledge. An internationally focused organisation, *Digital Intelligence (DQ) Institute*, works extensively across industry, not for profits and education, developing a framework fundamental to empowering citizens within the digital age through digital intelligence.

8.2.1 Digital intelligence

The DQ Institute provides a common framework (or 'global standards') for digital intelligence competencies that encompass the eight areas of digital life: digital literacy, digital use, safety, security, emotional intelligence, communication, rights, and identity (DQ Institute, 2023b). Based on the IEEE 3527.1 standard, the digital intelligence (DQ) Global Standards incorporate 24 digital competencies developed at three levels of citizenship, creativity, and competitiveness (Digital Intelligence, 2021). The specified knowledge, skills and attitudes/values outlined for each digital competency overlap behavioural and technical competencies, with knowledge and skills emphasising (but not exclusively) technical competencies and attitudes/values emphasising behavioural competencies. The DQ Framework encompasses OECD's Better Life Initiative (11 indicators of well-being) and the 17 UN Sustainable Development Goals and aims to 'guide digital practices towards achieving individual and societal wellbeing across all aspects of one's life' (DQ Institute, 2023b).

8.2.2 Children's needs missing from cybersecurity education frameworks

The DQ Institute positions itself as a comprehensive, collaborative platform, uniting various stakeholders, including funders, programme providers, and educators from various sectors, all dedicated to the shared objective of bridging the digital skills divide. This commitment is embodied in their DQ Global Standards, which are grounded in extensive international research and aim to establish a universal framework for digital competencies (DQ Institute, 2023a). Despite the Institute's holistic approach, it is noteworthy that a significant portion of these standards – about one-third – do not cater to the needs of children. This oversight is particularly striking given the recent introduction of the fourth iteration of the DQ Global Standards, titled 'Digital Connectivity', which, at first glance, appears to align more closely with the learning requirements of K-12 students. However, this alignment necessitates further scrutiny (DQ Institute, 2023b).

A critical examination of the Digital Competitiveness component of the DQ Global Standards reveals a language and framework that resonates more with professional environments than young learners' educational contexts. This discrepancy underscores a broader issue: the emphasis on adult digital competencies in the workforce while overlooking the specific digital skill sets necessary for children's development and safety online. Such oversight highlights a gap in the research and structured knowledge concerning digital education for the younger demographic and questions the inclusivity of global frameworks to close the digital skills gap.

The authors of this critique advocate for a more nuanced approach to digital education standards, one that recognises and addresses the unique requirements of children's digital literacy and safety. By shifting the focus from a predominantly adult-centric model to one that encompasses the educational needs of children, the goal is to contribute to a more comprehensive strategy for narrowing the digital skills gap on a global scale. This shift necessitates additional research and the reorganisation of existing knowledge to ensure that global standards are equally applicable and beneficial to the cyber educational needs of the younger population.

8.2.3 Terminology not yet relevant to students' needs

Internationally, 'vigilance' describes a specific core cybersecurity competency, featured in nearly half (43%) of research papers on cybersecurity education to date (Ibrahim, McKee, Sikos & Johnson, 2023) and referenced widely in industry. Variations of precise definitions exist across psychology and the computer science/cybersecurity fields. However, the authors have previously published an argument for vigilance to refer to 'sustained conscientious attention' from an interdisciplinary perspective (Ibrahim et al., 2023). Recent research in occupational health consistently points to vigilance as unsustainable in security technicians (for instance, Guidetti et al., 2023), who must stay alert to identify cyberattacks. In this context, the concern for any definition is that it is relevant to the needs of young learners. Children with their level of use of digital devices and services do not need to be alert in this way.

8.3 COMPETENCIES IN CYBERSECURITY EDUCATION

STEM and non-STEM disciplines differ significantly in terms of ontology and epistemology (Charlesworth & Banaji, 2019). This determines how students perceive computing-related subjects and, in turn, can substantially affect which educational approaches may effectively teach psychological principles to cybersecurity students (Taylor et al., 2017). Incorporating psychology into cybersecurity education requires an understanding of student motivations and expectations, as well as perspective-shifting because students tend to be interested more in the *how* and the *what* than the *why*, which prevents a deeper understanding of cyberattacks (Taylor-Jackson et al., 2020). Taylor et al. (2018) argue that 'to understand the potential risks of socio-technical systems, cybersecurity students need to understand and consider how people perceive, remember, feel, think, and solve problems' (p. 99).

Two main categories of cybersecurity competencies need to be developed in K-12 students: behavioural and technical competencies. They are equally important. We begin by discussing the most critical behavioural and essential technical competencies.

8.4 BEHAVIOURAL COMPETENCIES

Children learn cyber behavioural patterns from parents, siblings, teachers, and peers (Renaud & Prior, 2021). Understanding learners' online behaviour requires a multidisciplinary approach incorporating brain research, biology, psychology, human cognition, developmental theory, sociology, motivation theory, and social behaviour (Zheng, 2012).

8.4.1 Definitions

For this chapter, the authors reference behaviour as broadly inclusive of all the above-mentioned areas outlined by Zheng. Cognitive, meta-cognitive, and socio-emotional competencies are called 'behavioural competencies'. This tryptic division of competencies is used by DQ Institute, where they state that 'Digital Intelligence (DQ) was developed to encompass a comprehensive set of technical, cognitive, meta-cognitive, and socio-emotional competencies, which are grounded in universal moral values and enable individuals to face the challenges of digital life and adapt to its demands' (Digital Intelligence, 2021). The idea of 'universal moral values' being objective, delineates a strong line in the sand whereby the foundation of cybersecurity education (and especially for children) is as a 'lived ethic' aligned with shared community values, particularly privacy and freedom. This is as opposed to cybersecurity founded on, for instance, principles designed to protect a nation.

Behavioural cybersecurity competencies define what students' online secure and insecure actions are; examine the impact on an individual's well-being, health and safety; and delineate action (and non-action) that both protect internet users and promote safe and responsible use of technology (Türker & Çakmak, 2019).

8.4.2 Order of learning

Cybersecure behaviour is the result of cybersecurity awareness. Parsons et al. (2017) developed the *Human Aspects of Information Security Questionnaire (HAISQ)*, measuring information security awareness in employees based on the KAB learning model (knowledge–attitude–behaviour). This learning model claims, 'As an employee's knowledge of safe information security behaviours in the workplace is increased, their attitude improves, resulting in improved information security behaviour' (Parsons et al., 2017).

8.4.3 Teachers' modelling behaviour

Regarding learning, student competencies and awareness of online threats and privacy need to start with teachers modelling the behaviour, but currently, they are not (Hipsky & Younes, 2015). This situation is despite teachers in countries such as the United States being aware of, for instance, federal privacy policies.

8.4.4 Learning environments

Just like learning cybersecurity terminology, learning cybersecurity behaviours and competencies requires the flexibility to be activated within *nuanced, novel, and varied environments*. Scenario-based activities necessitate learners' critical thinking (a cognitive competency) across various environments and vigilance to pick up and act on various threats (the core behavioural competency) within different sets of circumstances. Scenario-based learning enables students to understand security through 'doing' security. This cross-relational capacity building activates learners' situational intelligence, a core behavioural competency the authors argue is required for digital literacy (Johnson et al., 2023).

A student's exposure to scenarios within their learning experiences propels individual learners towards self-directed cybersecure behaviour, with the transferability of these behavioural (or technical) skills from one domain to another. Students apply their improved knowledge of how their cybersecure thinking can be applied to many more new, unknown and unregulated digital environments and situations they find themselves in, and therefore act as their first line of defence.

8.4.5 Being aware of cyberthreats

While technical countermeasures, such as content filtering, can be applied to protect minors (Meeder, 2005), it is neither enforced to be used at schools or home nor can be considered 100% reliable, which is why being aware of cyberthreats is crucial. *Cyber awareness* is an established term with quality published reference points, furthered with arguments explaining how a term translates into the experience and behaviours of children or the cause and effect of cyberthreats on those experiences. Javidi and Sheybani (2018) reference the definition of cybersecurity awareness from the US National Institute of Standards and Technology (NIST) with a focus on teaching, describing it as 'teaching students about security concerns and threats and teaching them to respond accordingly' (p. 2). This focus may or may not encompass an industry definition; the example highlights how variations in the definition of key terms can be tricky at best when required to cross disciplines, as is the case between education and cybersecurity.

The development of safer online environments for children, and specifically, to identify and prevent online harassment and victimisation such as cyberbullying and cyberstalking, which occur regardless of socio-economic status or cultural background, requires not only technological tools but also policy and legislation reforms (Güllü et al., 2023; IRMA, 2020). Even contemporary online learning environments, such as those of massive open online courses, are prone to cyberbullying and cyber-mobbing (Grella et al., 2017).

8.4.6 Being able to apply best practices (for cyber hygiene)

As previously mentioned, *cyber hygiene* signifies the online practices required to stay secure, as referred to by a Canadian study (Witsenboer et al., 2022). Some of these practices are not trusting strangers online, not sharing sensitive information online (and not oversharing), prioritising secure activities over simplicity/convenience/time pressure, and being vigilant (paying attention to detail and knowing when to be suspicious, for example, being able to detect anomalies always to recognise spam). These can be taught to become second nature without necessarily knowing technical details. As addressed earlier, vigilance is overkill for the needs of children practising cybersecurity. Habit precedes vigilance, and the reoccurrence of experiences across different environments is the reinforcer towards best practice becoming second nature, for best practice becoming habitual.

Cyber wisdom, the ability to do the right thing at the right time when using computing devices and services, can be cultivated via formal education through four components: cyber wisdom literacy, cyber wisdom reasoning, cyber wisdom self-reflection, and cyber wisdom motivation. These can be built on character education literature and practice, such as understanding the nature of and applying virtues such as honesty, evaluating and prioritising these virtues, reflecting on moral dimensions, and acting online with the common good in mind (Polizzi & Harrison, 2022).

8.4.7 Being able to distinguish between ethical and unethical online behaviour

Ethics has been identified as a competency within organisational psychology (ethics competency), defined as 'the ability to apply knowledge, skills and abilities to incorporate values and principles that distinguish right from wrong when making decisions and choosing behaviours' (Cooper et al., 2017). As an example of its current application in schools, in the Australian curriculum, *ethical understanding* is a general capability taught across the different learning areas from K-12, introduced into the national curriculum in 2010. The papers collectively suggest examining how empathy as a learned state relates to ethics competency, a topic possibly covered in education and psychology, offering insights for cybersecurity education.

8.4.8 Not sharing personal data online

An essential behavioural competency that can be taught to young children is the ability not to share personal data online (or offline). First, understanding what comprises personal data is needed, and this information is unique to an individual. The ability to keep personal information private enables the prevention of more sophisticated applications of identity theft. Within this behavioural competency, we can consider appropriate personal information to share with particular people. Educating children about what is appropriate to share with a parent, compared to a teacher or a supermarket worker, may increase understanding about privacy protection. While 3- to 5-year-olds may not have their device, they will likely be temporal users of their parent's or carer's smartphone. The children must be aware of not only *not sharing* their personal information (such as name, age, where they live) but also realise that if they share their parent's or carer's phone number, that is sharing *someone else's* personal information, and that this is a breach of their privacy.

In many classrooms, it is a common yet suboptimal practice to share passwords for convenience and ease of use (for instance, leaving children's passwords in plain sight on the classroom wall – a cybersecurity nightmare). Two significant problems emerge from this practice. Firstly, the fact that classroom teachers engage in such insecure behaviour sets a negative example for their students. Secondly, the same credentials are connected to their accounts for an extended period (often years) without any systematic encouragement to change (unless a policy to do so is enforced via the school's infrastructure). School protocols and policies often drive behavioural change in such circumstances, particularly at the school administrative or governance levels.

One of the issues with behavioural-based competencies is that they are not simple to measure. The recent revision of the Australian national curriculum outlined many specific and explicit updates within the learning areas section

(Maths, etc.). The same level of detail does not apply to updating the general competencies section, which, at the broadest educational level from K to university, speaks to the development of the transferrable skills required for individuals and a cybersecure nation.

8.5 TECHNICAL COMPETENCIES

The following sections discuss the main technical competencies for cybersecurity education. From the educational perspective, human competencies are implicitly intertwined with technical competencies. Without comprehending these technical aspects, we cannot affect behavioural change nor achieve desired learning outcomes.

8.5.1 Having digital literacy

Various definitions of digital literacy have been refined over the years as the societal use of digital devices has increased. While *literacy* focuses on reading and writing, *digital literacy* includes accessing and using online information safely and appropriately and evaluating and creating content (Law et al., 2018).

Within previous versions of the national curriculum of Australia, the general capability was called *Information and Communication Technology* (ICT), which encompassed appropriate and effective use of digital technologies across all learning areas (ACARA, 2000), alongside investigating, communicating, and creating with ICT. This general capability is designed to be developed or applied within the context of learning areas (subjects). The latest Australian Curriculum version 9.0 retitled this general capability as *Digital Literacy* (ACARA, 2023a). Digital literacy includes investigating, communicating, and creating with digital tools but also incorporates the practice of digital safety and well-being, the management and protection of content, and the respect of intellectual property. Digital literacy is arguably a required technical competency of cybersecurity as the general capabilities are highlighted as equipping Australians to 'live and work successfully in the twenty-first century' (ACARA, 2023c). From a cybersecurity perspective, digital literacy is an essential technical competency to meet the needs of a cyber workforce and a cybersecure society, forming the basis for behavioural and other technical competencies to develop.

8.5.2 Being able to use private browsing

Web browsers allow excessive user tracking and profiling, actively used for social engineering. However, modern browsers provide a private browsing feature that blocks websites' attempts to track users with cookies, web beacons, etc. While web browsers are amongst the first software tools users get familiar with, many users use browsers with default, rather than strict, settings, which enables user tracking.

8.5.3 Being able to set up a strong password or passphrase

Even in the era of multi-factor authentication, passwords and passphrases continue to play their role in information security. While even beginner users regularly use (basic) passwords, often automatically assigned by their organisation, very few are familiar with complex passphrases and the best practices around these (Bitwarden, 2023). Despite being particularly important for children, this is because the password principles they learn early in their lives will be the ones they keep using later (Renaud & Prior, 2021).

8.5.4 Being able to utilise antivirus software

Understanding the importance and efficient use of antivirus software is imperative in cybersecurity education. This is because, despite the common misbelief, antivirus software remains vital in maintaining the security posture of computing devices.

8.5.5 Understanding where data is stored with cloud accounts and data ownership

Users focusing on the convenience of cloud services are often unaware of the potential threats of unsecured data transfer to and from the cloud, nor the question of data ownership, which might affect data use, alignment with company policies, and data retention. Once a cloud account is created, users might never be able to delete the account. In addition, it happens all too often that user data is replicated in near real time to multiple servers that users have no access to (only the service provider does), and in case the user initiates a file deletion, the service provider will still preserve a copy of the file, without the consent, let alone permission, of the user.

8.5.6 Being able to back up data

The rise of ransomware attacks in recent years necessitates a global understanding of the importance of regular data backups. When such attacks succeed, one of the best options is to restore computing systems, virtual machines, operating systems, and user files from the latest backup. This is not feasible without a robust backup infrastructure and proper configuration, regularity, occasional verification, and functional redundancy. Although students often save data on corporate (in-house, data centre, or cloud) infrastructure, there are many situations when they need to be able to back up their data themselves, such as when using their own devices.

8.6 THE SOCIOCULTURAL EFFECT OF HUMAN CYBERSECURITY COMPETENCIES

Society does not value cybersecurity (Johnson et al., 2023), instead seeing it as a 'problem for the IT department'. Nevertheless, parents and teachers are deeply concerned for the safety and security of young people in their care when on digital devices. Every day, the media includes horror stories of data breaches and cyberattacks whilst hearing and telling their own stories of deceptive text messages and phishing emails received daily. One of the critical components required in generating positive learning environments for students, teachers, and parents is the wider stance taken by society. As malintent across digital services occurs across the board for all of us, within the greater context of minimal national regulation for digital apps and other services, cybersecurity education for children cannot be considered in isolation. Again, we are all our first line of defence.

This chapter has addressed how cybersecurity knowledge indicates attitudes towards valuing cybersecurity. This erudition must be considered at the whole society level (and with the whole-of-school approach engaging and activating those secure learning environments) because a 'whole system initiative' (Johnson et al., 2023, p. 17) is 'needed across all industries and all workers to keep our identities and workplaces secure' (p. 17). Society needs a shift in attitude, preceded by cybersecurity awareness.

8.6.1 Addressing the workforce shortage

K-12 cybersecurity education is essential to teach primary and secondary school students knowledge and skills and prepare those interested in a cybersecurity career for post-secondary degree programmes offered in cybersecurity (Prickett et al., 2022).

The current shortage and future demand for skilled cybersecurity professionals is well-documented in Australia and internationally as both a trending and an emerging need (National Skills Commission, 2022). These trends are escalated by the ubiquitous nature of cybersecurity, itself being industry-agnostic, embedded within other industries, and has become an industry in its own right. Globally, we have companies that provide cybersecurity services across all sectors. Within all industry sectors, cybersecurity specialists operate in other industry-specific environments. Hence, career skills extend far beyond the more expected technical knowledge sets to also encompass psychology, organisational management, business development and strategy, administration, education and training, human resources, design, investigative services, communications, and marketing, to consider a few, as well as industry-particular knowledge paired with security knowledge. The skill sets required for cybersecurity to be integrated across sectors, including K-12 education (and then further integrated into everyone's daily activities), are tremendously broad.

8.6.2 Diversity is not in technology use: it is in people

The interdisciplinary nature of cyber-based roles accentuates the industry's need for diverse people with broad interests. These interests must be cultivated within educational settings to build a strong workforce, protect our economic sovereignty, and maintain the way of life people take for granted. By bringing *people* to the forefront of the image of cybersecurity, we align with the reality of cybersecurity not being simply technical. It is also behavioural. It is thinking, being, acting, and aligning with protective practices and ethical practices. The benefits of utilising adequate protections on everyday digital devices must be shown to bring about peace of mind, increase confidence, protect reputation, lessen worry, reduce stress, and ensure individuals, family, and friends do not lose their hard-earned money to threat actors.

8.6.3 Positive representation of cybersecurity

A societal standpoint of 'diverse people' being the public face of cybersecurity needs to be generated to move the current media portrayal of a 'tech-head' type of person who analyses data wearing a hoodie in a dark corner. Or is he the baddie? The media's misrepresentations perpetuate misunderstandings about cybersecurity in society. From a simple web search, the imagery used to represent cybersecurity is predominantly set in black backgrounds with hidden faces. This generates an air of secrecy, unfriendliness, and something alienating, or at least socially undesirable. For the average digital user, cybersecurity is socially alienating and complex to understand, even unobtainable. More engaging media representation should positively showcase a more comprehensive array of personalities.

8.6.4 Need for sociocultural approach to impact

Another part of the 'awareness' solution lies in addressing adults' or teachers' comprehension of the value of cybersecurity in our everyday lives (Johnson et al., 2023). We can borrow from our established understanding of privileged countries' occupational health and safety standards and protocols and align with existing societal-level learning curves. Hence, a media push generating a more positive portrayal that engages the humanity embedded within security is required to meet national goals worldwide. Highlighting diverse individuals with varied skills and backgrounds in different industry roles is crucial for societal change. A vital shift is needed in how the public and media portray the human-led cybersecurity skills essential to the workforce.

8.6.5 Shifting the mindset of a generation through a national conversation

Often shaped by media and governments, national conversations are crucial in enhancing online children's safety and security (Martin & Rice, 2012). Drawing parallels with historical workplace safety campaigns and other national safety efforts is sensible, given the similar need to protect children's interactions with digital platforms like websites, apps, text messaging, social media, and games. An example of such regulation in traditional media is the imposition of watershed broadcasting times on television, which limits content based on the viewer's age to ensure it is appropriate. However, digital content does not currently face similar stringent restrictions or guidelines. Online content regulation is primarily overlooked by parents, teachers, schools, and governments. For instance, there are no specific restrictions on advertising apps aimed at children, and while websites may ask for age confirmation, there is often no real verification of this information (O'Neill, 2013).

It is worth mentioning that some parents do make use of parental controls available through advanced antivirus software or streaming subscriptions to manage their children's screen time and content exposure. However, this is not widespread and does not address all potential risks. The lack of comprehensive oversight for digital content is partly attributed to the internet's inherent nature as a free and largely uncensored space.

The dialogue surrounding online child safety is essential in cultivating a societal mindset that values cyber safety and cybersecurity for children and holds content providers to the same standards of accountability expected in more traditional forms of media like television and radio. By elevating the public importance of cyberspace security, there is potential for a positive shift in how digital spaces are perceived and regulated. Notably, the strategy should recognise that preventative and restrictive measures might not be as effective as empowering children to make informed decisions using digital devices.

The human factor in cybersecurity education has its roots in the national psyche. Continuing with the 'roots' metaphor is the cybersecure tree of life, where tree branches are our attitude to online privacy and online ethics, and children's learning takes place at the leaves, able to grow and explore freely from a solid, secure foundational culture of positive cybersecurity practices and behaviours. This allows the tree – a nation, a world – to support and flourish, whilst children and those curious about cyberspace remain the first line of defence against the bad weather of malintent and unregulated soil conditions.

8.7 CONCLUSION

Cybersecurity is crucial to our existence in the digital world and integral to modern-day life. While the variety and sophistication of cyberthreats continue to rise, and children start using digital devices and services at a younger age than ever before, cybersecure practices and core cybersecurity countermeasures ought to be taught throughout K-12. Educational approaches in this discipline should consider both the technical aspects and the human competencies that can lead to positive behavioural changes in cyberspace. This is important not only to individuals but also to entire societies. Major educational approaches identify and develop skills and competencies via contemporary techniques, whether by injecting them into existing curricula (covering concepts in other subjects) or providing extracurricular activities, such as training courses and educational campaigns.

REFERENCES

ACARA. (2000). *Information and communication technology capability* (version 8.4). www.australiancurriculum.edu.au/f-10-curriculum/general-capabilities/information-and-communication-technology-ict-capability/

ACARA. (2023a). *Digital literacy*. https://v9.australiancurriculum.edu.au/f-10-curriculum/general-capabilities/digital-literacy?element=0&sub-element=0

ACARA. (2023b). *Digital technologies*. https://v9.australiancurriculum.edu.au/teacher-resources/understand-this-learning-area/technologies#digital-technologies

ACARA. (2023c). *General capabilities*. https://v9.australiancurriculum.edu.au/teacher-resources/understand-this-general-capability/digital-literacy

Bitwarden. (2023). *World Password Day Survey 2023*. https://bitwarden.com/resources/world-password-day/

Blažič, B. J., & Blažič, A. J. (2022). Cyber security skills among European high-school students: A new approach in the design of sustainable educational development in cyber security. *Sustainability*, *14*, 4763. https://doi.org/10.3390/su14084763

Charlesworth, T. E. S., & Banaji, M. R. (2019). Gender in science, technology, engineering, and mathematics: Issues, causes, solutions. *The Journal of Neuroscience*, *39*(37), 7228–7243. https://doi.org/10.1523/JNEUROSCI.0475-18.2019

Cooper, C. D., Hellriegel, D., & Slocum, J. W., Jr. (2017). Mastering organisational behaviour. Flatworld. ISBN 978-1-4533-7950-9.

De Luca Picione, R., Martini, E., Corona, F., Falzarano, M. L., & Cicchella, S. (2022). Bullying and cyber-bullying: A dialogue between psychology, sociology, and law to understand and to counteract youth violence. In E. Marchisio (Ed.), *Handbook of research on applying emerging technologies across multiple disciplines*. IGI Global. https://doi.org/10.4018/978-1-7998-8476-7.ch020

Digital Intelligence. (2021). *IEEE standard for digital intelligence (DQ) – Framework for digital literacy, skills, and readiness*. https://standards.ieee.org/ieee/3527.1/7589/

DQ Institute. (2023a). *Digital Intelligence Institute*. www.dqinstitute.org

DQ Institute. (2023b). *What is the DQ framework?* www.dqinstitute.org/global-standards/

Grella, C. T., Staubitz, T., Teusner, R., & Meinel, C. (2017). Can MOOCs support secondary education in computer science? In M. E. Auer, D. Guralnick, & J. Uhomoibhi (Eds.), *Interactive collaborative learning* (Vol. 1, pp. 478–493). Springer. https://doi.org/10.1007/978-3-319-50337-0_45

Guidetti, O. A., Speelman, C., & Bouhlas, P. (2023). A review of cyber vigilance tasks for network defence. *Frontiers Neuroergonomics, Sec. Cognitive Neuroergonomics*, *4*, https://doi.org/10.3389/fnrgo.2023.1104873; www.frontiersin.org/articles/10.3389/fnrgo.2023.1104873/full

Güllü, H., Karahan, E., & Akçay, A. O. (2023). A comprehensive investigation of cyberbullying and cyber victimisation among secondary school students. *Education and Information Technologies*, *28*, 12633–12650. https://doi.org/10.1007/s10639-023-11687-8

Hipsky, S., & Younes, W. (2015). Beyond concern: K-12 faculty and staff's perspectives on privacy topics and cyber safety. *International Journal of Information and Communication Technology Education*, *11*(4), 51–66. https://doi.org/10.4018/IJICTE.2015100104

Ibrahim, A., McKee, M., Sikos, L. F., & Johnson, N. F. (2024). A systematic review of K-12 cybersecurity education around the world. IEEE Access, 12, 59726-59738. https://doi.org/10.1109/ACCESS.2024.3393425.

IRMA. (2020). Developing safer online environments for children: Tools and policies for combatting cyber aggression. *IGI Global*. https://doi.org/10.4018/978-1-7998-1684-3

Javidi, G., & Sheybani, E. (2018). K-12 cyber security education, research, and outreach. In *2018 IEEE Frontiers in Education Conference*. IEEE. https://doi.org/10.1109/FIE.2018.8659021

Johnson, N. F., Ibrahim, A., Sikos, L. F., & Glowrey, C. (2022). *Cyber security curriculum in western Australian primary and secondary schools*. Interim Report: Curriculum Mapping. Report for the Cyber security Cooperative Research Centre and the Office of Digital Government, WA. https://doi.org/10.25958/x9r3-d254

Johnson, N. F., Ibrahim, A., Sikos, L. F., & McKee, M. (2023). *Going beyond: Cyber security curriculum in western Australian primary and secondary schools* (Final report). Report for the Cyber security Cooperative Research Centre and the Office of Digital Government, WA. https://doi.org/10.25958/41ZN-5R55

Law, N., Woo, D., Torre, J., & Wong, G. (2018). *A global framework of reference on digital literacy skills for indicator 4.4.2*. UNESCO Institute for Statistics, UNESCO Digital Literacy. https://unesdoc.unesco.org/ark:/48223/pf0000265403.locale=en

Lonergan, A., Moriarty, A., McNicholas, F., & Byrne, T. (2021). Cyberbullying and internet safety: A survey of child and adolescent mental health practitioners. *Irish Journal of Psychological Medicine*, *40*(1), 43–50. https://doi.org/10.1017/ipm.2021.63

Martin, N., & Rice, J. (2012). Children's cyber-safety and protection in Australia: An analysis of community stakeholder views. *Crime Prevention and Community Safety*, *14*, 165–181. https://doi.org/10.1057/cpcs.2012.4

Meeder, R. (2005). Access denied: Internet filtering software in K-12 classrooms. *TechTrends*, *49*(6), 56–78. https://doi.org/10.1007/BF02763732

Mihci Türker, P., & Kılıç Çakmak, E. (2019). An investigation of cyber wellness awareness: Turkey secondary school students, teachers, and parents. *Computers in the Schools*, *36*(4). https://doi.org/10.1080/07380569.2019.1677433

National Skills Commission, Australian Government. (2022). *State of Australia's skills 2021: Now and into the future: Digital skills in Australia and internationally*. www.jobsandskills.gov.au/

Olweus, D. (2012). Cyberbullying: An overrated phenomenon? *European Journal of Developmental Psychology*, *9*(5), 520–538. https://doi.org/10.1080/17405629.2012.682358

O'Neill, B. (2013). Who cares? Practical ethics and the problem of underage users on social networking sites. *Ethics and Information Technology*, *15*, 253–262. https://doi.org/10.1007/s10676-013-9331-4

Parsons, K., Calic, D., Pattinson, M., Butavicius, M., McCormac, A., & Zwaans, T. (2017). The human aspects of information security questionnaire (HAIS-Q): Two further validation studies. *Computers & Security*, *66*, 40–51. https://doi.org/10.1016/j.cose.2017.01.004

Polizzi, G., & Harrison, T. (2022). Wisdom in the digital age: A conceptual and practical framework for understanding and cultivating cyber-wisdom. *Ethics and Information Technology*, *24*, 16. https://doi.org/10.1007/s10676-022-09640-3

Prickett, T., Yang, L., Irons, A., Miller, K., Brooke, P., Crick, T., Hayes, A., Davenport, J. H., English, R., Maguire, J., Bechkoum, K., & Jones, A. (2022). Challenges and opportunities of teaching cyber security in UK university computing programmes. In L. F. Sikos & P. Haskell-Dowland (Eds.), *Cyber security teaching in higher education* (p. 8). Springer. https://doi.org/10.1007/978-3-031-24216-8_1

Renaud, K., & Prior, S. (2021). Children's password-related books: Efficacious, vexatious and incongruous. *Early Childhood Education Journal*, *49*, 387–400. https://doi.org/10.1007/s10643-020-01067-z

Taylor, J., McAlaney, J., Hodge, S., Thackray, H., Richardson, C., James, S., & Dale, J. (2017). Teaching psychological principles to cyber security students. In *2017 IEEE Global Engineering Education Conference*. IEEE. https://doi.org/10.1109/EDUCON.2017.7943091

Taylor, J., Thackray, H., Hodge, S. E., & McAlaney, J. (2018). Introducing psychological concepts and methods to cyber security students. In J. McAlaney, L. A. Frumkin, & V. Benson (Eds.), *Psychological and behavioural examinations in cyber security* (pp. 98–108). IGI Global. https://doi.org/10.4018/978-1-5225-4053-3.ch006

Taylor-Jackson, J., McAlaney, J., Foster, J. L., Bello, A., Maurushat, A., & Dale, J. (2020). Incorporating psychology into cyber security education: A pedagogical approach. In M. Bernhard, A. Bracciali, L. J. Camp, S. Matsuo, A. Maurushat, P. B. Rønne, & M. Sala (Eds.), *Financial cryptography and data security* (pp. 207–217). Springer. https://doi.org/10.1007/978-3-030-54455-3_15

Witsenboer, J. W. A., Sijtsma, K., & Scheele, F. (2022). Measuring cyber secure behaviour of elementary and high school students in the Netherlands. *Computers & Education*, *186*, Article 104536. https://doi.org/10.1016/j.compedu.2022.104536

Ybarra, M. L., Diener-West, M., & Leaf, P. J. (2007). Examining the overlap in internet harassment and school bullying: Implications for school intervention. *Journal of Adolescent Health*, *41*(6), S42– S50. https://doi.org/10.1016/j.jadohealth.2007.09.004

Zheng, R. Z. (2012). *Evolving psychological and educational perspectives on cyber behaviour*. IGI Global. https://doi.org/10.4018/978-1-4666-1858-9

Chapter 9

Incident response drills on cyber ranges

Enhancing cyber-defence education

Alexander E. Grojek, Leslie F. Sikos, David J. Holmes, and Oliver Guidetti

9.1 INTRODUCTION

The emergence of a new wave of generative AI-powered and constantly growing plethora of cyberthreats necessitates efficient teaching and training of cybersecurity analysts at an increasing pace. The cybersecurity world faces an ever-growing number of complex and dynamic threats. As a result, it has become essential to train future cybersecurity professionals on comprehensive technology environments rather than relying on a limited set of tools, practices, methods, and approaches.

The traditional approach to cybersecurity education often struggles to provide students with a practical, hands-on experience due to difficulties in securing industry placements, and the unavailability of authentic datasets hinders realistic training scenarios. Developing case studies authentically replicating the complexity of a real-world incident is daunting. In response, cyber ranges emerge as a transformative solution. These controlled and interactive environments allow participants to engage in simulated cyber incidents, providing a hands-on learning experience that bridges the gap between theoretical knowledge and practical application.

Cyber ranges create a dynamic space for experimental learning by simulating real-world cyber incidents in a controlled environment. Participants can actively respond to scenarios such as malware outbreaks, data breaches, and network intrusions, allowing them to develop and refine their incident response skills. Notably, the controlled nature of these simulations enables educators and trainers to monitor participants' performance and decision-making processes. This monitoring aspect facilitates continuous improvement, allowing for targeted feedback and personalised guidance to enhance the participants' skills over time.

Proactive incident response requires a solid technical background in communication networks and computer security and hands-on skills that rely on experience that could only be obtained via work placements that often cannot be arranged or working on authentic case studies (Sikos, 2023a). Such case studies cannot rely solely on static datasets because these would

 DOI: 10.1201/9781032664859-9

not provide authentic experience during training; the pressure, and in the case of threat hunters, the thrill, typically associated with working on cyber-incidents promptly, is lost, and there is no live network traffic. In addition, security analysts need to be trained in teamwork as well, such as in the form of a green team (infrastructure operators), a white team (organisers), a red team (attackers), and a blue team (defenders).

These issues face many challenges, from obtaining resources to ensuring that production sites and live services are unaffected on the network. Moreover, educators might find it challenging to repeat successful *ad hoc* training events or classes (Vykopal et al., 2017). One way to overcome these challenges is to use a live but controlled network environment on a *cyber range* (named after police and military shooting ranges used for hands-on practice), such as the *CyberBit Range* (CyberBit, 2020; Sikos, 2023b). Some cyber ranges may be available as a *Cyber Range as a Service (CRaaS)* (Angelogianni et al., 2021).

9.2 CYBER-DEFENCE EDUCATION AND THE NEED FOR PRACTICAL TRAINING

Cyber ranges are pivotal as sophisticated technical platforms that serve a dual purpose: enabling cutting-edge research and development initiatives while providing a dynamic training and exercise environment within the cyber domain (Karjalainen & Kokkonen, 2020). At their core, cyber ranges emulate essential networks and systems, establishing a versatile foundation to support these activities, consisting of a closed and tightly controlled environment equipped with the necessary systems, tools, and network infrastructure. A distinguishing feature of cyber ranges is their ability to replicate realistic internet simulations, encompassing the generation of background traffic and user emulation. This comprehensive simulation extends beyond the mere imitation of cyber activities; it strives to mirror the intricacies of the real-world cyber landscape. This emulation includes but is not limited to scenarios involving sophisticated attacks, mimicking cyber adversaries' tactics, techniques, and procedures. By doing so, cyber ranges provide a nuanced and authentic setting for researchers, professionals, and trainees to engage with genuine cyberthreats.

The cyberthreat landscape constantly evolves, with cybercriminals becoming more sophisticated and innovative in their attacks. These threats include malware, phishing attacks, ransomware, and DDoS attacks. Threat actors range from individual hackers to well-funded organised crime groups and nation-states. With the rise of IoT devices, cloud computing, and interconnected systems, the attack surface has expanded, making it more challenging to defend against cyberthreats. The escalating complexity of network systems has intensified the occurrence of unexpected behaviours and dependencies

(Karjalainen & Kokkonen, 2020). This complexity underscores the importance of understanding the events in the cyber domain, assessing the statuses of valuable assets, and comprehending the impacts of various dependencies on these assets. In this intricate threat landscape, achieving situational awareness becomes crucial, as it plays a pivotal role in the cyber domain, allowing for a comprehensive understanding of the dynamic cyber environment.

Cyber-defence education is crucial in producing competent professionals to combat these evolving threats. With proper education and training, individuals can develop network security, cryptography, ethical hacking, incident response, and secure software development skills. Cybersecurity professionals are essential for protecting sensitive data and critical infrastructure and ensuring the overall stability of the digital economy. One of the challenges in cybersecurity education is the gap between theoretical knowledge and practical skills and the dynamic nature of information infrastructure settings (Arshad et al., 2021). Traditional academic programmes often focus on theoretical concepts, while real-world cyberthreats require hands-on skills and practical experience. Bridging this gap is essential to producing professionals who can effectively mitigate cyberthreats. Hands-on training, simulated environments, and practical exercises are crucial to developing the skills to defend against real cyberattacks.

Cyber ranges are specialised training platforms designed to simulate real-world cybersecurity scenarios. These platforms provide a controlled environment where individuals can practice and improve their cybersecurity skills without the risk of damaging live systems and may also be provisioned as a part of a smart campus (Tian et al., 2018). Cyber ranges allow users to engage in realistic exercises, including defending against simulated cyberattacks, conducting penetration testing, and responding to security incidents. Cyber ranges allow individuals to gain hands-on experience and develop practical skills in a safe and controlled environment. These platforms enable learners to apply their theoretical knowledge, enhance problem-solving abilities, and understand the complexities of managing and securing various systems and networks. Cyber ranges are valuable tools for educational institutions, cybersecurity professionals, and organisations seeking to train and assess their workforce's capabilities in the ever-changing cyberthreat landscape.

The closed nature of cyber ranges is a crucial attribute that sets them apart from other educational aids. This closed environment offers a controlled and secure space, isolating the simulated cyber activities from external networks. This isolation is critical, as it ensures the confidentiality of often sensitive research and development efforts and creates a risk-free environmental setting for conducting training exercises. In this controlled environment, professionals can deploy and test realistic cybersecurity threat scenarios without fear of unintended consequences or actual harm to operational systems and networks. Cyber ranges are not confined to theoretical simulations; they can actively incorporate genuine attacks and intrusions into their scenarios. This hands-on approach supports cybersecurity professionals and researchers in addressing real-world challenges in a controlled setting, honing their skills and developing effective strategies for countering evolving threats. By offering a risk-free yet authentic environment, cyber ranges become invaluable tools for enhancing theoretical understanding and practical skills to navigate the complex and ever-evolving cyber landscape.

The *Defence Advanced Research Projects Agency (DARPA)* played a pioneering role in establishing one of the earliest cyber ranges, acknowledging the advancing frontiers of cybersecurity and the critical need for research and development fostered through rigorous testing and experimentation (DARPA, 2024; Karjalainen & Kokkonen, 2020). The National Cyber Range (NCR) initial iteration was a foundational prototype DARPA developed. Over time, researchers have adopted NCR as a paradigm for implementing cyber ranges. This pivotal contribution has spurred the global development of diverse cyber ranges by entities, including government, industry, and universities to research centres and national security organisations. These cyber ranges display a broad spectrum of capabilities, from modest laboratory-based single-server testbed infrastructures to elaborate, virtualised internet-like frameworks. Beyond their training utility, cyber ranges have proven highly effective for research and development endeavours.

Moreover, the Cyber Defence Pooling & Sharing project of the *European Defence Agency* (EDA) has recognised the necessity for cooperation between national cyber ranges at the European level (European Defence Agency, 2024; Karjalainen & Kokkonen, 2020). This initiative acknowledges the different capabilities of various cyber ranges and the growing demand for simulating the complexity of the cyber domain. Cyber ranges are designed to cater to specific needs and audiences, making their capabilities diverse. They are used in different scenarios, such as training individuals or groups, including employees of companies, organisations, and governments – the heterogeneous nature of cyber ranges results from their purposeful construction (Päijänen et al., 2021). Cyber ranges also serve additional applications from cybersecurity research and development, certifying products or services, conducting competence assessment, and even recruitment activities. The multifaceted role of cyber ranges underscores the necessity to simulate a comprehensive and intricate cyber environment, accounting for unforeseen dependencies and their potential consequences.

9.3 THE ROLE OF CYBER RANGES IN CYBERSECURITY EDUCATION

A cyber range is a controlled, virtual environment specifically designed for simulating real-world cybersecurity scenarios and exercises, allowing the highest level of cyber-training experience. It serves as a training platform where individuals, including students, cybersecurity professionals, and organisations, can practice and enhance their skills in a secure setting. Cyber ranges mimic various systems, including computer networks, servers, routers, firewalls, and operating systems (such as Windows, Linux, and macOS). These virtual environments can replicate diverse network architectures, from small local area networks (LANs) to complex enterprise-level infrastructures.

In terms of hardware and software components, a cyber range typically consists of powerful servers and network equipment to host the virtualised systems and create realistic network topologies. Virtualisation software, such as *VMware* or *Hyper-V*, creates and manages virtual machines representing different devices and systems. Security tools and applications, including firewalls, intrusion detection systems, and vulnerability scanners, are integrated to replicate the functions of real-world cybersecurity solutions (Cooley et al., 2022; VMWare, 2024). Additionally, cyber ranges often incorporate simulation software that generates lifelike cyberthreats and attack scenarios, allowing participants to respond to incidents, conduct penetration testing, and practice defensive strategies. These components provide an immersive and practical learning experience within a controlled digital environment, enabling individuals to develop and refine their cybersecurity skills effectively. Cyber range platforms are highly customisable and scalable and offer a multitude of capabilities that can be utilised in teaching and learning:

- Support for and replication of enterprise-grade ICT infrastructures covering various platforms (Windows, Linux), as well as industry control system (ICT)/operational technology (OT) environments, is enhanced by automatic scenario emulators. These emulators can generate benign/legitimate/background network traffic – this enables scenario emulation repeatability, which in turn can be measured and allows for the reliable comparison of teams across various days' sessions. Additionally, the emulators can execute a preconfigured attack, making red teams optional in the range setup or conducting live injection during simulation sessions.
- Off-the-shelf content: simulation technology plus content with increasing levels of difficulty.
- Complex cyber-incidents with technical details, in-depth scenario documentation, and orchestration for event management.
- Scenario composer for repurposing existing and creating new cyber-incident cases via leveraging a library of commercial systems and applications and collaboratively replicating complex IT/OT infrastructures.
- Wearable sensors worn by individuals can provide valuable insight into the pressure of the participants throughout the training sessions.
- Instructor feedback: real-time comments, session debriefing with videos of entire simulation sessions, individual and team performance, and decision scoring.

One of the requirements for preparing efficient incident response plans is staff training (Karie & Sikos, 2022). While cyber ranges can be used for conducting complex incident detection and response exercises, the dynamic nature of real-world systems and the adaptability for changing requirements would require *task automation* in the cybersecurity exercise life cycle (Yamin & Katt, 2022), including *automating training content generation* (Beuran et al., 2018). This needs to be complemented by the monitoring of training participants. For this reason, Andreolini et al. (2020) proposed a *distributed monitoring architecture* that gathers relevant information about trainees' activities during cyber range exercises. This architecture aims to (1) measure the speed at which a user is progressing towards an objective and (2) assess the correctness of the actions taken.

The learners' roles and target knowledge, skills, and abilities must be covered when using cyber ranges for training. A prime example of this is the *Cyber Range Design Framework*, proposed by Katsantonis et al. (2023), which adopts both modern and traditional learning theories, facilitates element reuse and allows dynamic assessments. It adapts to the learners' needs and capabilities and provides real-time feedback.

The cognitive complexity (C) of cyber range event operating environments (O) can be modelled as the weighted arithmetic mean of the quantified properties $p_1 \ldots {}_{nn} \in P$ (e.g., variety of traffic types and paths of traffic flow, variety of user behaviours, etc.) as given by Equation 9.1.

Cognitive complexity of cyber range event operating environments:

$$C(O) = \frac{\sum_{i=1}^{n} w_i p_i}{\sum_{i=1}^{n} w_i} \tag{9.1}$$

where $w_1 \ldots {}_n \in W$ represent weights that describe the impact of each property p on the final cognitive complexity measure, whether low, medium, or high (Roque et al., 2019). Note that task performance cannot be used directly to measure this complexity.

Cyber ranges serve various use cases, encompassing security research, testing, development, and certification efforts within the cybersecurity domain (Päijänen et al., 2021). The concept of a development testbed predates the currently known concept of a cyber range, with development teams creating these environments to assess the functionality of their updates in a setting resembling the production environment. Concurrently, research environments strive for authenticity, often employing isolated ICT environments to investigate malware without internet connectivity. In the domain of security education, there is a correlated relationship between competence building and assessment, where certification bodies provide cyber ranges as practical environments for students seeking validation of their skills. This approach positions the cyber range environment as a viable option for assessing students' evolving competence. The evolution of an individual's expertise is integral to the broader development of cyber capabilities and resilience, as viewed from an organisational perspective.

Cyber ranges are vital in bridging the gap between theoretical knowledge and practical skills and holistically teaching how to see cyberattacks and defence mechanisms. Traditional education often imparts a theoretical understanding of cybersecurity concepts, but practical application is essential in this rapidly evolving field. In addition, authentic case studies rely on complex networks, potentially with hundreds of participants and LANs (Liljenstam et al., 2005). Cyber ranges provide a platform where individuals can translate theoretical knowledge into hands-on experience, reinforcing their understanding of concepts and techniques. By confronting real-time threats and challenges, learners gain a deeper appreciation for the complexities of cybersecurity, allowing them to bridge the divide between theory and practice effectively. This experiential learning approach ensures that cybersecurity professionals are well-equipped to tackle the dynamic and diverse landscape of cyberthreats in the modern digital age. The main types of cyber ranges used in education include the following:

- *Simulation ranges:* Recreating a synthetic network environment mimicking the behaviour of real network components
- *Emulation ranges: Running on dedicated network infrastructure*, encompassing the mapping of network, server, and storage infrastructures onto physical infrastructure
- *Overlay ranges:* Running on top of real networks, servers, and storage infrastructure
- *Hybrid ranges:* A custom combination of any of the above

Purpose-designed multi-level training platforms have also been proposed, which cover hands-on and theoretical challenges, scenario templates, and capabilities for testing and evaluation, monitoring, recording, and scoring (Subaşu et al., 2017). Educators can utilise cyber ranges within both pre-packaged and *ad hoc* curricula in the implementation of core as well as advanced incident detection material developed by either professional or educational communities like K-SEC (Yonemura et al., 2022) and individual training of incident detection and response tasks mimicking the functions of a security operation centre. Operations related to ad hoc situations can be used to teach vulnerability assessment and ethical hacking. Tutors can use cyber ranges in alignment with education frameworks, such as the *National Initiative for Cybersecurity Education (NICE)* (National Institute of Standards and Technology, 2023).

Thrilling live jeopardy-style capture-the-flag (CTF) competitions are suitable not only for extracurricular activities but also in a class environment with students alternating between offensive and defensive roles (Katsantonis et al., 2017). CTFs also allow the creation of virtual machines to be used for formative assessments (Chothia & Novakovic, 2015).

9.4 ENHANCING PSYCHOLOGICAL READINESS AND RESILIENCE WITH CYBER RANGES

A fundamental difference between training and education must be recognised in cybersecurity education (Lunn et al., 2021). Technical skills are paramount in the education of cybersecurity analysts; however, the psychological preparedness of individuals to respond to cyber emergencies is often overlooked (Lunn et al., 2021). Education concerns underlying principles, theories and technical skills required to defend a network. In contrast, training extends education and equips learners with the operational preparedness to respond to cyber emergencies (Lunn et al., 2021). By analogy, commercial pilots must complete *in vivo* training as a matter of safety, not just classroom education in operating an aircraft (Todd & Thomas, 2013). This competency-based approach could benefit cybersecurity professionals (Somarakis et al., 2019). For example, cybersecurity professionals typically receive a foundational level of education through undergraduate or master's degrees. However, to develop specialised skills in defending critical infrastructure from types of cyberattacks, they can benefit from training in the simulated environments of cyber ranges. These environments replicate real-world scenarios, such as a port or telecommunications grid, allowing operators to become operationally prepared to tackle specific types of attacks (Somarakis et al., 2019).

Cyber range training also formalises continued professional development (CPD) requirements for cybersecurity operators. Formalised CPD requirements uplift the professional attributes of a profession, ensure operators' skills are kept up to date, and expand beyond the curriculum of university degrees (Psychology Board of The Australian Health Practitioner Regulation Agency, 2019). For example, clinical psychologists in Australia must fulfil CPD requirements annually to maintain their registration with the *Australian Health Practitioner Regulation Agency* (AHPRA, 2024). Engaging in regular simulated emergency incidents on cyber ranges would likewise enhance the professional capacity of cybersecurity operators by exposing them to the intensity and pressures of real-world cyberthreats in a safe and controlled environment (Knox et al., 2019).

This exposure is crucial for building the mental resilience of cybersecurity analysts, enabling them to maintain composure better and think critically during emergency incident response. By repeatedly facing simulated threats, participants develop a stress-resistant mindset, enabling them to handle real-world scenarios more effectively (Knox et al., 2019). Indeed, specialised sectors such as maritime cybersecurity have adopted this competency-based approach. For example, CyberMAR (2023) is a federated cyber range that enables security professionals to simulate complex attack–defence scenarios on port infrastructure (Amditis, 2017). The scenarios are immersive experiences reflecting real-world maritime infrastructure demands. These simulations ensure that the professionals using them for training are theoretically knowledgeable and practically adept at handling threats and situations unique to the maritime sector. By focusing on practical skills and real-time decision-making, cyber ranges such as *Cyber-MAR* can aid in developing a workforce that is well-prepared and resilient against the evolving landscape of highly specialised cyberthreats to maritime infrastructures.

CyberMAR (2023) is one example of a range cybersecurity operators can use to enhance their professional capabilities. However, as cybersecurity expands within the Australian context, there will be an increasing need for cyber ranges that specialise in different classes of the *Security of Critical Infrastructure Act* (SOCI, 2018). For example, the Register of Critical Assets, defined in the SOCI Act (2018), comprehensively covers various cyber infrastructure classes. This extensive list includes the following:

1. *Communications Networks:* Vital for maintaining nationwide and global connectivity
2. *Energy Systems:* Essential for power generation and distribution
3. *Financial Systems:* The backbone of economic transactions and stability
4. *Healthcare Systems:* Critical for patient care and medical data management
5. *Transport Networks:* Encompassing both passenger and freight movement
6. *Water and Wastewater Systems:* Fundamental for public health and sanitation
7. *Government Databases:* Containing sensitive and essential public records
8. *Electoral Systems:* Crucial for maintaining democratic processes
9. *Public Service Delivery Systems:* Key for government service provision
10. *Industrial and Manufacturing Systems:* Central to the production and supply chain
11. *Emergency Services:* Vital for public safety and crisis response

Additionally, the act specifies various assets within these categories:

- Broadcasting Assets (Section 12E)
- Domain Name Systems (Section 12KA)
- Data Storage or Processing Assets (Section 12F)
- Freight Services Assets (Section 12C)
- Water and Wastewater Assets (Section 5)
- Hospitals (Section 5)
- Food and Grocery Assets (Section 12K)
- Energy Market Operator Assets (Section 5)

- Freight Infrastructure Assets (Section 12B)
- Liquid Fuel Assets (Section 12A)
- Public Transport Assets (Section 5)
- Maritime Ports (Section 11)
- Gas Assets (Section 12)
- Financial Market Infrastructures and Payment System Assets (Section 12D)

The detailed classification of critical infrastructure under the SOCI Act (Australian Department of Home Affairs, 2018) underscores each category's unique and complex nature, presenting distinct security challenges. This intricate landscape necessitates the development of an industry of specialised cyber ranges that serve cybersecurity professionals to train effectively in safeguarding these diverse and vital environments (Australian Department of Home Affairs, 2018).

The Australian Government's 2023–2030 Cybersecurity Strategy (2023) calls for a whole-of-nation effort to uplift Australia into the most cyber-secure nation by 2030. Achieving this objective necessitates a concerted, nationwide effort that formalises CPD requirements for cybersecurity professionals, particularly those working with SOCI-registered critical assets. This strategy emphasises the importance of formalising CPD requirements for these professionals, extending beyond their initial educational qualifications. This initiative will require the development of a dedicated training industry of cyber ranges that caters specifically to the needs of the cybersecurity sector. This approach ensures that cybersecurity professionals' skills are updated regularly, keeps pace with the emerging threats in the virtual landscape, and ensures a baseline level of regulated skill that other professions, such as clinical psychology and aviation, have already adopted. Such continuous skill enhancement is critical for bolstering Australia's capability to protect and defend its critical cyber infrastructure effectively and meeting the goal of becoming the most cybersecure nation on Earth by 2030.

9.5 ADVANTAGES OF INCIDENT RESPONSE DRILLS ON CYBER RANGES

Cyber ranges offer immersive learning experiences by simulating real-time cyber incidents. Participants can engage with lifelike scenarios, such as malware infections, data breaches, and DDoS attacks, allowing them to respond to these incidents as they would in the real world. This immersive approach provides a hands-on understanding of the complexities of handling actual cyberthreats, enhancing participants' skills and confidence in dealing with similar situations in their professional roles. The authenticity of the case studies investigated on a cyber range can motivate students, and the experience gained can provide value both in their learning journey during candidature and in their career pathways after graduation. Beyond hyper-realistic attack scenarios, such as ransomware attacks, DoS attacks, SCADA attacks (factory silent attack, overload, or shutdown), or website defacements, students get the opportunity to master the following:

- Corporate network environments
- Critical infrastructure/SCADA security
- Advanced cyberthreats (e.g., advanced persistent threat (APT), privilege escalation, DDoS SYN flood, data exfiltration, man in the middle)
- Contemporary cybersecurity toolkits covering next-generation firewalls, IPS, and complex Security Operations Centres (SOCs)/Security Event Information Management Systems (SIEMs)/Security Orchestration, Automation, and Response (SOAR) solutions
- Complex skills: incident detection and response (including incident response playbooks), network forensics, reverse engineering
- Efficient SOC teamwork, potentially as a part of a capstone unit/applied project

Having a proven record and demonstrable hands-on skills in responding to incidents involving industry systems simulated in cyber ranges may also increase employer demand for graduates. In a cyber range environment, participants must make decisions under pressure, mirroring the high-stakes situations in actual cyber incidents. This fosters the development of critical thinking skills as individuals must analyse rapidly evolving situations, assess risks, and choose effective response strategies. Practising critical decision-making in a controlled setting prepares cybersecurity professionals to handle stressful scenarios with composure and precision, ensuring a more effective response to real cyberthreats. Working in high-pressure environments requires making timely decisions and ensuring the accuracy of the actions taken to achieve the best possible outcome in a given situation. Delayed good decisions, as well as poor or incorrect decisions, can have severe consequences and side effects.

Cybersecurity incidents often require collaboration and effective communication among team members. Cyber ranges facilitate team-based response exercises where participants analyse, strategise, and execute incident response plans. By practising collaborative problem-solving and clear communication, individuals learn how to coordinate efforts, share vital information, and optimise their collective expertise. These teamwork skills are an invaluable asset in a real-world environment, where a meaningful response to cyberthreats often necessitates timely collaboration between different cyber specialists within an organisation. Note that making good decisions in detecting cyberthreats and mitigating cyberattacks does not simply rely on background knowledge and technical skills but also depends on the contextual information available to the response team at that moment, which influences team adaptation (Deline et al., 2021), and on trust, risk behaviour (Granåsen & Andersson, 2016), and cognitive bias (Pfleeger & Caputo, 2013).

Cyber ranges allow organisations to test and refine their incident response plans. By simulating diverse cyber incidents, teams can assess the effectiveness of their existing plans, identify weaknesses, and make necessary improvements. Regular testing in a cyber range environment ensures that incident response plans are up to date, relevant, and well-coordinated, enabling organisations to respond swiftly and efficiently when actual incidents occur.

The dynamic nature of cyberthreats requires cybersecurity professionals to be adaptable and agile. Cyber ranges expose participants to various scenarios, challenging them to adapt their strategies to different types of attacks. By encountering diverse challenges and learning to adjust their responses accordingly, individuals become more versatile and better equipped to handle evolving cyberthreats. Nurturing adaptable and agile cyber defenders through cyber range training ensures they can effectively counteract new and sophisticated cyberthreats as they emerge, safeguarding digital assets and systems with resilience and expertise.

9.6 PARALLELS TO AVIATION TRAINING

This section explores the utilisation of flight simulators in training pilots and aviation crew to unveil the benefits of this technology in a highly regulated industry. The focus is on drawing parallels to flight simulator training technologies and techniques that may be adaptable for incorporation into the cyber range domain. Hence, it is proposed that the cybersecurity community consider a focus on the aviation domain to extract valuable insights and knowledge from an industry with both a more extended history and a more highly regulated operational framework yet sharing similarities of purpose. The rewards of such a strategy are manifold, encompassing cost-efficiency, scenario repeatability, domain-specific familiarisation, risk mitigation, and adaptive training, all contributing to a comprehensive and meaningful evaluation of personnel performance and capabilities.

The adoption of flight simulators for civil aviation training and performance evaluation has matured to such a degree that they have become an essential and mandatory device for the effective and efficient training of aircrews (Lofaro & Smith, 2023). The now pervasive practice of using flight simulators within the civil aviation domain results from the appreciation of the heightened aviation skill set derived from such simulators and a convergence of other influences. In its early developmental stage, and continuing into the early 1990s, civil aviation relied upon a pool of experienced ex-military personnel as a source of aviation trainers and advisors. This resulted in the ex-military personnel directly influencing the increased use of the flight simulator in training civil aircrew. This resulted from the familiarity of the ex-military personnel with the use of flight simulators in military aircrew training programmes.

The inception of flight simulators dates back to 1929 when the first rudimentary simulator replicated the cockpit and functionalities of a flight deck. Generic in nature and designed for general aviation only, it quickly developed into a military flight trainer during the Second World War. The military use of flight simulator technology was a primary driver of early flight simulator technology. As aviation technology developed, military aircraft rapidly became more complicated and technology driven. The flow-on effect was that civil air transport began to demand aircraft manufacturers that the aircraft's flight simulator accompany each new aircraft delivery. As the fidelity of these formative simulators continued to improve, many airlines began to appreciate the benefits of flight simulation technology in terms of both safety and financial return (Allerton, 2000). This development was supported through performance and innovation in computer technology, with the simulator manufacturers able to meet the exacting standards necessary to comply with increasing legislative obligations and to simulate aircraft dynamics more accurately in reflection of that experienced in the cockpit (Allerton & Lammertse, 2023). These advances culminated in several significant trends.

Firstly, as domestic computer performance improved, so did the availability and adoption of commercial off-the-shelf (COTS) flight simulator systems. Previously, simulators were bespoke devices with a price point only available to large and financially well-resourced companies. Secondly, a marked divergence of fidelity between simulator manufacturer products was developed. The subsequent flow-on effect introduced an undesirable disparity in training methods and delivery used in aircrew training. To address this divergence between simulator products, strict regulations were introduced by authorities to mandate simulator quality consistency. There was a particular focus to ensure that simulators more accurately modelled the aircraft they were supposed to simulate and more accurately monitored the skills assessments conducted by airlines to confirm that the simulator training provided met rigorous and objective standards. As a result, by the 1990s, international consensus had been reached such that aviation stakeholders adopted aviation regulations in most countries. For example, today, we have civil aviation flight simulators for the B-757/767 that cost over $35 million and ongoing maintenance and operational costs of thousands of dollars per hour (Lofaro & Smith, 2023).

Through the confluence of contingent vectors, the period from the late 1970s until today has witnessed the evolution of the flight simulator as an indispensable tool for aviation training and certification.

9.6.1 Flight simulator: a comprehensive definition and functional purpose

The essence of a flight simulator can be comprehensively defined by its fundamental purpose, which serves as the cornerstone for its design and application. Firstly, a flight simulator aims to meticulously present simulated information, ensuring that it faithfully replicates the intricacies of a real-world operational system (Allerton & Lammertse, 2023). This involves the recreation of the aircraft's dynamics, systems, and environmental conditions to an extent that mirrors the challenges and nuances encountered in actual flight. The simulator becomes a dynamic, virtual realm where pilots can interact with simulated aircraft and respond to various scenarios, fostering an environment rich in realism.

Secondly, the flight simulator is designed to accommodate, actively facilitate, and enhance the aircrew's technical and cognitive skills (Allerton & Lammertse, 2023). This goes beyond mere replication; it involves creating scenarios challenging and refining the skills required for proficient piloting. The simulator is a dynamic training ground where pilots can hone their decision-making abilities, emergency response skills, and overall aeronautical expertise in a controlled yet realistic setting.

Crucially, the overarching goal of these simulator-driven skill enhancements is to catalyse 'training transfer'. This term encapsulates the seamless transition of acquired skills and knowledge from the simulated environment to the real-world operational context. Through targeted and realistic training scenarios, the flight simulator contributes to developing aircrew with improved functional capability. The skills acquired and refined within the simulated space are intended to directly translate into enhanced performance and adaptability when operating in the complexities of the actual aviation environment.

The flight simulator is a nexus between theoretical knowledge and practical application, functioning as a replication tool and a dynamic training platform. By aligning its purpose with the facilitation of training transfer, the flight simulator becomes an invaluable asset in cultivating aircrew proficiency, ultimately contributing to broader aviation safety goals, efficiency, and pilot competence.

9.7 THE HUMAN FACTORS CHALLENGE IN TRAINING TRANSFER

Understanding the human factors problem in training transfer is crucial to optimise the efficiency and effectiveness of training programmes across all industries, including cybersecurity. The extent of the transfer problem is subject to varying estimates, reflecting the complexity of ensuring that trained individuals can successfully apply their acquired skills in the real-world context. Georgenson (1982) estimated a conservative 10% success rate in achieving behavioural change post-training. In contrast, Saks' (2002) survey data depicts a more concerning notion, indicating that approximately 40% of students fail to transfer immediately after training. Moreover, this data reveals a progressive decline in transfer success, with 70% faltering after one year and only 50% of training investments resulting in tangible organisational or individual improvements.

These statistics underscore a critical concern: the considerable discrepancy between training funding and its return on investment. This misalignment raises questions about the efficiency of current training methodologies and necessitates a revaluation of human resource training efforts. To address this, interventions must be focused on enhancing the individual, the team, and organisational human resource performance (Yamnill & McLean, 2001). Flight simulators must be crucial in these interventions as a pivotal tool in human factor training.

However, predicting the transfer of training in advance, especially within flight simulation, poses a considerable challenge. It is exceedingly difficult to anticipate how well skills acquired in a simulated environment translate to real-world scenarios. A proposed method to objectively assess the transfer of training involves conducting tests with two distinct groups of trainees. One group exclusively receives airborne training, while the other undergoes flight simulator training. By comparing the competency levels between these two cohorts, it becomes possible to gauge the effectiveness of each training method.

This testing hypothesis rests on the assumption that specific tasks can be measured objectively within each learning environment. Allerton and Ross (1991) emphasise the importance of developing clear and measurable criteria for evaluating competency in airborne and simulator-based training. This approach offers a pragmatic strategy for addressing the human factors problem in training transfer, fostering a more nuanced understanding of how skills acquired in a simulated setting align with real-world proficiency. Ultimately, overcoming the human factors challenges in training transfer requires thoughtful integration of research, assessment methods, and target interventions to enhance the overall impact of training efforts on individual and organisational performance.

9.7.1 Application of training transfer

Training transfer, in the context of skill acquisition and application in real-world settings, involves effectively integrating learned knowledge and abilities into the practical environment. This process necessitates the seamless transition of acquired skills from the training setting to the workplace, where they must be consistently applied and sustained over an extended period. As stated by Baldwin and Ford (1988), for successful training transfer to occur, 'Learned behaviour must be transferred to the workplace environment and subsequently maintained there over an extended period' (p. 63).

In the aviation industry, particularly in training pilots, the concept of training transfer is exemplified through flight simulators. These sophisticated tools allow for establishing clear objectives and measurable standards of competency. For instance, achieving proficiency in flying an instrument approach under conditions of low or no visibility is a critical skill for pilots, which traditionally may require 25 hours of actual flight training. This training is time-consuming and can be fraught with risk, especially under adverse weather conditions. Flight simulators offer a safe and controlled environment where pilots can repeatedly practice these approaches, honing their skills without the associated risks of actual flights. A significant portion of this specialised training can be effectively offloaded to a flight simulator to enhance cost-effectiveness and reduce the inherent dangers of authentic flight training. This approach ensures pilots are well-prepared for challenging flight conditions and optimises training resources and safety.

Using flight simulators in pilot aviation training offers many advantages, including skill acquisition, cost-effectiveness, and risk mitigation. By providing a synthetically realistic and controlled environment, flight simulators enable students to develop skills without exposure to the inherent risks associated with actual flight. The immersive nature of flight simulators contributes to enhanced skill retention, allowing pilots to review and then revisit to reinforce their skillsets periodically. This approach aids in mitigating the effects of skill decay over time.

In essence, training transfer facilitated through meaningful simulator training contributes to developing more competent, proficient, and confident pilots. The culmination of these benefits ultimately leads to heightened aviation safety. The ability to effectively transfer and maintain skills acquired in a simulated environment to real-world scenarios underscores the significance of integrating innovative training methodologies in industries where precision, competence, and safety are paramount.

9.8 CYBERSECURITY AND AVIATION: TRAINING COMMONALITIES

Educational principles within cybersecurity and aviation domains share commonalities, highlighting the necessity for a comprehensive teaching concept addressing diverse subjects across multiple disciplines (Langner et al., 2022). Like cybersecurity, aviation training aims to enable learners, in this case, aspiring pilots, to realise their full potential through a blend of practical and theoretical education. Both domains employ various teaching methods to effectively prepare learners for the unique challenges within their respective domains. Additionally, the adaptability of teaching resources and materials, as emphasised in the cybersecurity context, aligns seamlessly with the dynamic nature of aviation training.

Formulating a comprehensive teaching concept for the cybersecurity discipline necessitates considering various requisites and factors from educators and students, alongside specific references to individual subjects and disciplines (Langner et al., 2022). The goal is to empower students to attain their full potential by delivering a practical and seamless integrated theoretical education through diverse methods to address cybersecurity challenges effectively. This approach enables educators to integrate and adapt established strategies while expanding their education frameworks of teaching resources, materials, and assignments by acquiring skills and competencies unattainable through conventional methods.

The efficacy of the learning process depends on factors directly and indirectly influenced by both educators and students, shaping the applied teaching methods during the learning process. Critical factors for successful learning in both the cybersecurity and aviation domains were identified (Langner et al., 2022).

9.8.1 Cognitive skills

Cognitive skills form the foundation of teaching in both domains, facilitating knowledge acquisition, information processing, logical thinking, and problem-solving. In cybersecurity, these skills are crucial to learning, decision-making, and attentiveness. In aviation training, cognitive skills are likewise crucial for pilots to acquire and process aeronautical knowledge, make informed decisions, and demonstrate logical thinking during flight operations.

9.8.2 Non-cognitive skills

Soft skills such as communication, cooperation, commitment, self-efficacy, and conscientiousness are essential in both domains. Beyond interpersonal aspects, they contribute to effective leadership and stress management. These skills are crucial for effective crew resource management, communication with air traffic control, and maintaining situational awareness during flights.

9.8.3 Social and compromise skills

The ability to interact and compromise socially is vital in both domains. In cybersecurity, these skills encompass social intelligence, interpersonal characteristics, verbal and non-verbal elements, and social self-regulation. They are essential for understanding complex topics and finding solutions, with adaptable, collaborative spaces addressing the significant educational challenge. In aviation, effective communication between cockpit crew members and ground personnel and the ability to compromise and interact socially is essential for achieving objectives and ensuring safe flight operations.

9.8.4 Learning strategies

Action patterns regulating an individual's learning categorised as cognitive strategies, repetitive learning, development, and organisational strategies are accommodated in both domains. This empowers individuals to optimise their educational objectives. Like cybersecurity professionals, pilots employ various learning strategies to regulate their learning, reflecting the need for continual development, review, and maintenance of their aeronautical knowledge.

9.8.5 Motivation and expectations

Acknowledging diverse learner motivations and expectations, teaching methods recognise that mismatches between methods and learner preferences can diminish motivation. Explicit course descriptions become pivotal to managing expectations and supporting motivated individuals through self-directed methodologies, fostering a conducive learning environment. These concepts align closely with those of aviation training in maintaining the motivation of aspiring pilots.

The critical significance of practical and theoretical education is evident in cybersecurity and pilot aviation training. In cybersecurity, precritical education involves hands-on experience with security tools, simulated attacks, and real-world scenarios, ensuring learners can effectively apply theoretical knowledge to practical situations. Similarly, in pilot aviation training, aspiring pilots undergo practical flight simulations and hands-on experience in aircraft, translating the theoretical aeronautical principles into real-world flying skills. Integrating theory and practical application enhances understanding and proficiency in both domains.

Furthermore, developing cognitive and non-cognitive skills stands out as a critical priority in both domains. In cybersecurity, cultivating cognitive skills such as critical thinking, problem-solving, and decision-making is imperative for addressing complex cyberthreats. Non-cognitive skills, including effective communication, collaboration, and stress management, are equally crucial. Similarly, in pilot aviation training, cognitive skills are essential for a pilot's ability to process information rapidly, make split-second decisions, and ensure the aircraft's and its passengers' safety. Non-cognitive skills, such as effective communication within the cockpit and with air traffic control, contribute to the overall success and safety of flight operations.

The alignment between teaching concepts and training factors in cybersecurity and pilot aviation training goes beyond surface-level similarities. It reflects a shared recognition of the multidimensional nature of learning, emphasising not only the transfer of knowledge but also the practical application of skills, adaptability to evolving circumstances, and the holistic development of cognitive and non-cognitive abilities. This synergy contributes to the overall effectiveness of educational strategies in preparing individuals for success in the intricate and dynamic domains of cybersecurity and aviation.

9.8.6 Challenges and considerations

One of the challenges in implementing incident response drills on cyber ranges is the presence of resource constraints and specific infrastructure requirements. Setting up and maintaining a cyber range involves significant financial investment in terms of hardware, software, and skilled personnel. Educational institutions and organisations may face limitations in funding and resources, making it difficult to establish and sustain fully equipped cyber ranges. Additionally, ensuring scalability to accommodate a growing number of participants and evolving technologies adds complexity to resource management.

Cyberthreats and cyberattack techniques are constantly evolving, making it essential for incident response drills to stay current and relevant. Attackers continuously develop new methods to exploit vulnerabilities and bypass security measures. Cyber ranges need to adapt by incorporating the latest attack vectors, malware strains, and tactics used by threat actors. This requires regular updates to the simulation scenarios and training modules, ensuring that participants are exposed to realistic and up-to-date challenges. Staying ahead of emerging threats is crucial to preparing cybersecurity professionals for the ever-changing landscape of cyberattacks.

When designing realistic training scenarios on cyber ranges, it is imperative to address ethical considerations. Simulated incidents should not involve illegal or unethical activities, even within a controlled environment. Maintaining a balance between realism and ethical boundaries is crucial to avoid crossing legal or moral lines. Educators and trainers must carefully craft scenarios challenging participants' skills without encouraging malicious behaviour or actions violating privacy, legal regulations, or ethical norms. Ensuring a robust ethical framework within training scenarios is essential for the credibility and integrity of cyber range-based education programmes.

Measuring the effectiveness of incident response drills conducted on cyber ranges is essential to assess the training's learning outcomes and overall impact. Traditional indicators include response time, accuracy in identifying and mitigating threats, and adherence to established protocols. However, assessing the participants' ability to apply learned skills in real-world situations is equally important. Surveys, participant feedback, and performance evaluations can provide qualitative insights into the effectiveness of the training programme. Additionally, tracking participants' progress over time and correlating their performance in drills with actual incidents, when applicable, can offer a comprehensive understanding of the training's effectiveness. Regular evaluation and feedback mechanisms are essential for refining training methodologies and ensuring that incident response drills on cyber ranges deliver meaningful educational outcomes.

9.8.7 Future implications and trends

Integrating artificial intelligence (AI) and machine learning (ML) technologies in cyber ranges revolutionises cybersecurity training. AI and ML algorithms can create dynamic and adaptive cyber scenarios based on participants' actions and responses, offering personalised and challenging learning experiences. These technologies can simulate sophisticated cyberthreats, analyse participants' behaviour, and provide real-time feedback and guidance. AI-driven cyber ranges enhance the training environment by mimicking evolving attack techniques, enabling learners to practice defending against AI-generated threats, and fostering a deeper understanding of AI and ML in cybersecurity applications.

Gamification techniques, such as point systems, leader boards, and rewards, can significantly enhance engagement and motivation in cyber ranges (Jelo & Helebrandt, 2022). By transforming training exercises into interactive games, participants are more likely to immerse themselves in learning. Moreover, incorporating diverse and creative scenarios, ranging from ransomware attacks to social engineering schemes, ensures learners are exposed to various challenges. Gamification and scenario diversity make training enjoyable and provide a realistic and engaging platform for honing practical skills, encouraging healthy competition, and promoting continuous learning.

Cyber ranges facilitate global collaboration in cyber-defence education, enabling students and professionals from diverse geographical locations to collaborate, share knowledge, and engage in joint training exercises. Virtual collaboration platforms within cyber ranges allow participants to work together in real time, fostering a sense of community and encouraging the exchange of best practices. Collaborative international cyber-range training initiatives enhance cultural understanding, promote information sharing, and create a unified front against global cyberthreats. Such collaborative efforts strengthen nations' and organisations' collective cybersecurity capabilities worldwide.

Anticipated advancements in cybersecurity technologies, including quantum computing, blockchain, and secure multi-party computation, are poised to transform the cyber-defence landscape. Quantum computing may break existing encryption methods, prompting the development of quantum-resistant algorithms and cryptographic techniques. Blockchain technology offers tamper-proof and transparent transaction records, revolutionising cybersecurity practices such as secure identity management and supply chain integrity. Secure multi-party computation enables sensitive data analysis without exposing the raw data, ensuring privacy in collaborative cybersecurity efforts.

These advancements, coupled with AI-driven threat detection, real-time response automation, and enhanced user authentication mechanisms, will reshape the way cyber defence is approached. However, they also bring new challenges, requiring cybersecurity professionals to continuously update their skills and adapt to emerging technologies. As the cyber-defence landscape evolves, staying ahead of these advancements is crucial to effectively protect digital assets and maintain the integrity and confidentiality of sensitive information. Education and training programmes, particularly those utilising advanced cyber ranges, will be pivotal in preparing the cybersecurity workforce to leverage these advancements for proactive and robust cyber-defence strategies.

9.9 CONCLUSION

The pivotal role of cyber ranges in cybersecurity education cannot be overstated, as they blend authenticity (complexity, dynamic nature) and practicality (as a platform providing options for teaching, monitoring/feedback, and assessment) to cultivate a comprehensive learning experience. The authenticity of cyber ranges, marked by their complexity and dynamic nature, mirrors the real-world challenges faced in cybersecurity. Simultaneously, their practicality emerges as a versatile platform offering educators and students a spectrum of teaching options, robust monitoring and feedback mechanisms, and practical assessment tools.

Crucially, the contribution of cyber ranges extends beyond technical proficiency, providing a bridge to the diverse and ever-evolving landscape of cyberthreats. As the idea of cyber-graduate employability requires training on an endless variety of systems not available to educators and students, cyber ranges provide a viable option for memorable hands-on experiences in SOC/SIEM-like environments. The lessons learned during cyber range sessions are invaluable and go far beyond technical skills, covering teamwork in high-pressure environments, insight into psychological traits of insider attacks such as data exfiltration, the diversity of tools typically used in incident detection and response, and more.

REFERENCES

Allerton, D. J. (2000). Flight simulation – Past, present and future. *The Aeronautical Journal*, *104*(1042), 651–663. https://doi.org/10.1017/S0001924000096901

Allerton, D. J., & Lammertse, P. (2023). Lessons learnt and lessons missed from flight simulation. *European Journal of Dental Education*, 1–8. https://doi.org/10.1111/eje.12919

Allerton, D. J., & Ross, M. (1991). Evaluation of a part-task trainer for ab initio pilot training. In *Training transfer RAeS conference*. https://scholar.google.com/scholar?q=Allerton,+D.J.+and+Ross,+M.,+Evaluation+of+a+Part-task+Trainer+for+Ab+Initio+Pilot+Training.+RAeS+Conference+Training+Transfer,+1991,+London,+UK

Amditis, A. (2017, November). *Cyber-MAR: Cyber preparedness actions for a holistic approach and awareness raising in the maritime logistics supply chain*. www.cyber-mar.eu/

Andreolini, M., Colacino, V. G., Colajanni, M., & Marchetti, M. (2020). A framework for the evaluation of trainee performance in cyber range exercises. *Mobile Networks and Applications*, *25*, 236–247. https://doi.org/10.1007/s11036-019-01442-0

Angelogianni, A., Politis, I., Polvanesi, P. L., Pastor, A., & Xenakis, C. (2021). Unveiling the user requirements of a cyber range for 5G security testing and training. In *26th international workshop on computer aided modelling and design of communication links and networks*. IEEE. https://doi.org/10.1109/CAMAD52502.2021.9617776

Arshad, S., Alam, M., Al-Kuwari, S., & Khan, M. H. A. (2021). Attack specification language: Domain specific language for dynamic training in cyber range. In *2021 IEEE global engineering education conference*. IEEE. https://doi.org/10.1109/EDUCON46332.2021.9454094

Australian Department of Home Affairs. (2018). *Security of critical infrastructure act 2018* (No. 29, 2018, Compilation No. 4). Federal Register of Legislation. www.legislation.gov.au/Details/C2022C00160

Australian Government. (2023). *2023–2030 Australian cyber security strategy*. www.homeaffairs.gov.au/reports-and-pubs/files/2023-2030_australian_cyber_security_strategy_discussion_paper.pdf

Australian Health Practitioner Regulation Agency. (2024). www.ahpra.gov.au

Baldwin, T. T., & Ford, J. K. (1988). Transfer of training: A review and directions for future research. *Personnel Psychology*, *41*, 63–105. http://dx.doi.org/10.1111/j.1744-6570.1988.tb00632.x

Beuran, R., Tang, D., Pham, C., Chinen, K., Tan, Y., & Shinoda, Y. (2018). Integrated framework for hands-on cyber security training: CyTrONE. *Computers & Security*, *78*, 43–59. https://doi.org/10.1016/j.cose.2018.06.001

Chothia, T., & Novakovic, C. (2015). An offline capture the flag-style virtual machine and an assessment of its value for cyber security education. In *The 2015 USENIX summit on gaming, games, and gamification in security education*. www.cs.bham.ac.uk/~tpc/SecEduVM/BhamEduVM.pdf

Cooley, S., Briggs, M., Heuer, D., Jahiu, D., Myers, A., Wilhite, C. D., & Lohr, H. (2022, April 26). *Hyper-V on windows 10* (Microsoft learn). https://learn.microsoft.com/en-us/virtualization/hyper-v-on-windows/

CyberBit. (2020). https://go.cyberbit.com/cyber_security_training-platform/

CyberMAR. (2023, February 7). Cyber-MAR: Cyber preparedness actions for a holistic approach and awareness raising in the MARitime logistics supply chain. www.cyber-mar.eu

DARPA. (2024). *Defence Advanced Research Projects Agency*. www.darpa.mil

Deline, S., Guillet, L., Rauffet, P., & Guérin, C. (2021). Team cognition in a cyber defence context: Focus on social support behaviours. *Cognition, Technology & Work*, *23*, 51–63. https://doi.org/10.1007/s10111-019-00614-y

European Defence Agency. (2024). https://eda.europa.eu

Georgenson, D. L. (1982). The problem of transfer calls for partnership. *Training & Development Journal*, *36*(10), 75–77. https://api.semanticscholar.org/CorpusID:222306397

Granåsen, M., & Andersson, D. (2016). Measuring team effectiveness in cyber-defence exercises: A cross-disciplinary case study. *Cognition, Technology & Work*, *18*, 121–143. https://doi.org/10.1007/s10111-015-0350-2

Jelo, M., & Helebrandt, P. (2022). Gamification of cyber ranges in cyber security education. In *20th international conference on emerging eLearning technologies and applications*. IEEE. https://doi.org/10.1109/ICETA57911.2022.9974714

Karie, N. M., & Sikos, L. F. (2022). Cyber security incident response in the enterprise. In M. Ahmed, N. Moustafa, A. Barkat, & P. Haskell-Dowland (Eds.), *Next-generation enterprise security and governance* (pp. 83–120). CRC Press. https://doi.org/10.1201/9781003121541-4

Karjalainen, M., & Kokkonen, T. (2020). Comprehensive cyber arena; The next generation cyber range. In *2020, the IEEE European Symposium on Security and Privacy Workshops*. IEEE. https://doi.org/10.1109/EuroSPW51379.2020.00011

Katsantonis, M., Panayotis, F., & Ioannis, M. (2017). Conceptual analysis of cyber security education based on live competitions. In *2017 IEEE Global Engineering Education Conference*. IEEE. https://doi.org/10.1109/EDUCON.2017.7942934

Katsantonis, M. N., Manikas, A., Mavridis, I., & Gritzalis, D. (2023). Cyber range design framework for cyber security education and training. *International Journal of Information Security*, *22*, 1005–1027. https://doi.org/10.1007/s10207-023-00680-4

Knox, B. J., Lugo, R. G., & Sütterlin, S. (2019). Cognisance as a human factor in military cyber defence education. *IFAC-PapersOnLine*, *52*(19), 163–168. https://doi.org/10.1016/j.ifacol.2019.12.168

Langner, G., Skopik, F., Furnell, S., & Quirchmayr, G. (2022). A tailored model for cyber security education utilising a cyber range. In *8th International Conference on Information Systems Security and Privacy* (pp. 365–377). https://doi.org/10.5220/0010834000003120

Liljenstam, M., Liu, J., Nicol, D., Yuan, Y., Yan, G., & Grier, C. (2005). RINSE: The real-time immersive network simulation environment for network security exercises. In *Workshop on principles of advanced and distributed simulation*. IEEE. https://doi.org/10.1109/PADS.2005.23

Lofaro, R. J., & Smith, K. M. (2023). Civil aviation: Flight simulators and training. In *Human factors in simulation and training: Application and practice*. https://doi.org/10.1201/9781003401353-3

Lunn, S., Ross, M., & Liu, J. (2021, December). Cyber security integration: Deploying critical infrastructure security and resilience topics into the undergraduate curriculum. In *2021 International Conference on Computational Science and Computational Intelligence (CSCI)* (pp. 866–871). IEEE. https://doi.org/10.1109/csci54926.2021.00203

National Institute of Standards and Technology. (2023, July 21). *NICE*. Author. www.nist.gov/itl/applied-/nice

Päijänen, J., Saharinen, K., Salonen, J., Sipola, T., Vykopal, J., & Kokkonen, T. (2021). *Cyber range: Preparing for crisis or something just for technical people?* Academic Conferences International Limited. https://doi.org/10.34190/EWS.21.012

Pfleeger, S. L., & Caputo, D. D. (2013). Leveraging behavioural science to mitigate cyber security risk. *Computers & Security*, *31*(4), 597–611. https://doi.org/10.1016/j.cose.2011.12.010

Psychology Board of the Australian Health Practitioner Regulation Agency. (2019, September 12). *Continuing professional development*. www.psychologyboard.gov.au/Registration/Continuing-Professional-Development.aspx

Roque, A., Stetson, D., & Hannon, D. J. (2019). Assessing the cognitive complexity of cyber range environments. *The Journal of Defence Modelling and Simulation: Applications, Methodology, Technology*, 17(1), 39–46. https://doi.org/10.1177/15485129188206

Saks, A. M. (2002). So what is a good transfer of training estimate? A reply to Fitzpatrick [dataset]. In *PsycEXTRA Dataset*. American Psychological Association (APA). https://doi.org/10.1037/e576922011-004

Sikos, L. F. (2023a, April 25). Exploring the use of synthetic data in international security. *United Nations Interview*.

Sikos, L. F. (2023b). Learning environments for digital forensics teaching in higher education. In L. F. Sikos & P. Haskell-Dowland (Eds.), *Cyber security teaching in higher education* (pp. 127–139). Springer. https://doi.org/10.1007/978-3-031-24216-8_6

Somarakis, I., Smyrlis, M., Fysarakis, K., & Spanoudakis, G. (2019, September). Model-driven cyber range training: A cyber security assurance perspective. In *International workshop on information and operational technology security systems* (pp. 172–184). Springer International Publishing. https://doi.org/10.1007/978-3-030-42051-2_12

Subaşu, G., Roşu, L., & Bădoi, I. (2017). Modelling and simulation architecture for training in cyber defence education. In *9th International Conference on Electronics, Computers and Artificial Intelligence*. IEEE. https://doi.org/10.1109/ECAI.2017.8166396

Tian, Z., Cui, Y., An, L., Su, S., Yin, X., Yin, L., & Cui, X. (2018). A real-time correlation of host-level events in cyber range service for smart campus. *IEEE Access*, 6, 35355–35364. https://doi.org/10.1109/ACCESS.2018.2846590

Todd, M. A., & Thomas, M. J. (2013). Experience, competence, or syllabus? Influences on flight hours at licensing of commercial pilots. *The International*

Journal of Aviation Psychology, *23*(2), 169–180. https://doi.org/10.1080/10508414.2013.772851

VMWare. (2024). *VMware by broadcom*. Author. www.vmware.com

Vykopal, J., Vizvary, M., Oslejsek, R., Celeda, P., & Tovarnak, D. (2017). Lessons learned from complex hands-on defense exercises in a cyber range. In *2017 IEEE Frontiers in Education Conference*. IEEE. https://doi.org/10.1109/FIE.2017.8190713

Yamin, M. M., & Katt, B. (2022). Modeling and executing cyber security exercise scenarios in cyber ranges. *Computers & Security*, *116*, 102635. https://doi.org/10.1016/j.cose.2022.102635

Yamnill, S., & McLean, G. N. (2001). Theories supporting transfer of training. *Human Resource Development Quarterly*, *12*(2), 195–208. https://doi.org/10.1002/hrdq.7

Yonemura, K., Oyama, S., Kobayashi, H., Yamada, S., Shiraishi, K., Izumi, S., Fukuda, T., Okamoto, H., Hirano, M., Fujimoto, Y., Moriyama, H., Sakamoto, Y., Sato, J., Noguchi, K., Taketani, H., & Kishimoto, S. (2022). Teaching expert development project by KOSEN security educational community. In *2022 IEEE Global Engineering Education Conference (EDUCON)*. IEEE. https://doi.org/10.1109/EDUCON52537.2022.9766560

Index

M

N

O

P

Q

R

S

T